I would like to thank MIT's Sloan School of Management for providing much needed data and analytics as well as the Harvard Kennedy School and Northwestern University's Kellogg School of Business for access to their online research archives.

Copyright © 2019 James L. Toland

Third Edition, October 21st, 2024

I0495414

The Paradox of Disruptive Technology. © Copyright 2019 James L. Toland

All rights reserved. No part of this publication to be reproduced, distributed, or transmitted in any form or by any means including photocopying, recording, or other electronic or mechanical methods, without the prior written permission of the author except in the case of brief quotations embodied in critical reviews and certain other noncommercial uses permitted by copyright law. For permission request, write to author, addressed "Attention Permissions Coordinator," at the address(es) below.

Library of Congress Control Number: 0000000000 (Pending)

First Publication 2019 by Son Serralta Publishing, USA

Author Contact: jauma23@gmail.com

If you enjoyed this read, please take a minute and write a lovely review.

Preface	3
Chapter 1- Enter The Jedi	13
Chapter 2 - Just Enough To Be Dangerous	22
Chapter 3 - The Future Is Here	27
Chapter 4 - Intelligent Effort	34
Chapter 5 - The Paradox of Disruptive Technology	39
Chapter 6 - Proof is In The Pudding	43
Chapter 7 - March On And Fear Not The Thorns	47
Chapter 8 - You're Gonna Need A Bigger Boat	51
Chapter 9 - If Your Gospel Isn't Touching Others, It Hasn't Touched You	59
Chapter 10 - Fall Seven Times And Stand Up Eight	68
2024 Appendix	72
1: Always Say Please and Thank You To your AI, or Else…	75
2: Ledger Technology - The Oldest Tech In The World	81
3: Hack This MF!	86
4: Web3 or Bust! (and Why Burning Man is So 20 Minutes Ago)	91
Glossary of Terms	96
About The Author	100

PREFACE

Working as an executive for technology companies over the years has been an immensely fulfilling career. I am passionate about bringing disruptive technology to market and take great delight in being the 'smartest person in the room' when it comes to rethinking how to leverage emerging business technology. It takes a certain sense of humor to call yourself the 'smartest person in the room' in Silicon Valley. If you are not familiar with the term, the title of this book may come off as profoundly arrogant. I assure you it is a business term widely used across many industries that simply connotes someone is a true expert in a particular field. A title bears little distinction as sometimes a freshman analyst might just be the resource tapped to provide critical data to the executive team. It is typically not the ultimate decision-maker or project lead, as they are the ones seeking expert counsel. Here is an extreme example, The President of the United States and the Joint Chiefs of Staff et al are in the situation room deliberating over a military strike on a terrorist training facility. The mission requires air support and a Naval force to complete the mission successfully. You may think the Joint Chiefs heading the Air Force and Navy would be the smartest people in the room, but you would be wrong. The Admiral of the Navy would be singularly the smartest person in the room because the Department of The Navy commands air, sea and land forces, also known as the U.S. Marines. If a plane takes off from an aircraft carrier it is most often a Navy pilot in a Naval Air Support Command aircraft. The Air Force typically would support a Naval air attack with long- range reconnaissance aircraft or long-range bombers from the mainland. In order for the expression 'smartest person in the room ' to have any meaning it has to be a singular person not a committee.

Mastermind Talks founder and networking guru Jayson Gaignard has built a career quoting the philosopher Confucius, who said *"if you are the smartest person in the room, you are in the wrong room."* In Gaignard's book <u>Mastermind Dinners: Building Lifelong Relationships</u> he explains that if you think you are the smartest person in the room than you are likely not open to new ideas from others equally as smart. He's not wrong. In fact, I have rarely met a C-level tech executive who didn't practice this kind of humility, save one, and his company went bankrupt. Gaignard's networking events teach how to use collaboration and community to build great technology. On its face this is not the worst approach to business problem solving. But at some point, in the exercise, you are going to have to hear from a few experts and those people are, for that moment in time, the smartest people in the room on the subject at hand. Confucius' wisdom is far older than the widely used business terminology so it's logical to presume that the term 'smartest person in the room' was coined not in contrast to the father of Eastern philosophy, but rather in support of his humble belief: micro vs. macro mastery of a subject.. It is interesting to note that a century after Confucius wrote the Five Classics, Socrates (who would not have known of Confucius) uttered the very foundation of Western philosophy, 'ipse se nihil scire id unum sciat', which simply means, 'all that I know for sure is that I know nothing at all.' If leadership goes into a project with this healthy attitude, then it is your job as the Account Executive (AE) to enlighten them with what you do know all the while being mindful and transparent as to what you do not know.

In tech, we are always looking for the next unicorn, the next whiz kid, the next great innovation. It's not so much about intelligence as it is brilliance. You might not be the owner of the legendary pseudonym Satoshi Nakamoto (Bitcoin pioneer and cryptocurrency servant), but if you have a *black box* technology that trades alt coins on Coinbase with a documented ROI of 17% month over month, you can stand tall in a bank's boardroom and exclaim with confidence, 'I am the smartest person in the room when it comes to cryptocurrency trading strategies.' This person may also be the kind of genius who fills his coffee cup to the brim and then is perplexed when they can't get enough cream in the cup. Here's a fun story: I was at a party years ago with a few rocket scientists from NASA's Jet Propulsion Laboratory at the gathering. I watched this Cal Tech PhD fill his first glass of the night with rum and coke but couldn't figure out how to add ice to the glass because it was too full of liquid. He struggled with this problem and then not so cleverly dumped some of his beverage into the sink. I found this to be incredibly shortsighted, but I can tell you that guy was the smartest person in the room when it came to rocket guidance systems; that's for damn sure. I use this example to illustrate that the phrase is not meant literally, but rather describes the undisputed authority on a particular subject.

I've aimed to keep the spirit of this guide light and approachable. As you very well know the wild, weird world of tech is pretty laidback and, for the most part, an industry where informal corporate cultures prevail. This guide is intended to be read in a wry humorous voice. I would also like to encourage you to keep an open mind when it comes to my strategic steps towards successfully selling disruptive tech. No two deals are the same and every technology solution comes with its own challenges. Some Software as a Service (SaaS) products are ready out of the box and don't require Operating Systems (OS) integrators, while others are full custom developments that can take well over a year to fully deploy. After all these years I would rather have quick integration paced deals that I can 'land and expand' rather than twelve-month-long projects involving brigades of developers. With that said I have addressed both kinds of scenarios.

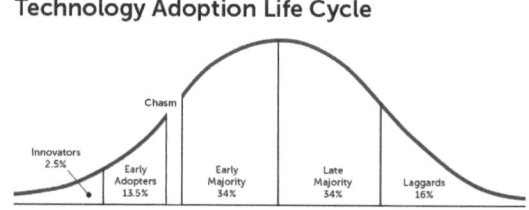

Geoffrey Moore's bestselling book <u>Crossing the Chasm: Marketing and Selling Disruptive Products to Mainstream Customers</u> is considered the bible for strategizing how to take emerging tech to market and was an early inspiration for my own sales acumen. Moore details how companies must implement unique strategies for marketing before and after what he refers to as 'the chasm'—the flatlining of sales between early adopters and early majority markets. Startups often struggle during this period in their revenue growth because early and late majority adopters are much harder to convince to adopt disruptive tech. This guide is a practical, step-by-step resource for salespersons interested in effective ways to pitch and win deals on both sides of the chasm.[1]

While effective marketing strategies help define your product's position in the marketplace, it takes a strategically-minded salesperson to successfully execute all steps of a disruptive technology sales cycle—from first pitch to testing, onboarding, and expansion. Let this guide be your practical resource to help you:

1) Identify markets

2) Create strategic sales plans

3) Sell with confidence

4) Effectively manage full sales cycles

5) Become a known Rockstar in your industry

It's one thing to adopt new platforms and/or software solutions to aid in processes, but to completely rethink one's processes takes great courage. There are many examples of this today: Bromium, Apple, MongoDB, VMware, Google, Amazon, and

---

[1] Image Source: http://www.themarketingstudent.com/wp-content/uploads/2017/04/chasm-adoption-lifecycle.jpeg

Wozniak's old company, Fusion-io. Who are the next unicorns that will replace these giants? It's not a matter of if but when. Since the dawn of innovation, technologies have been improved upon and eventually replaced by new inspirations that are always initially viewed as heretical and dangerous. How do we bring these products to market successfully today, when the majority of the world is still getting used to the old way of doing things? I've coined this **The Paradox of Disruptive Technology** ©.

There is a profound difference between innovation and disruption when it comes to defining emerging tech. Harvard Business School professor Clayton M. Christensen first coined the term 'disruptive technology' in his 1997 book <u>The Innovator's Dilemma</u>. Christensen categorizes emerging technology into two types: sustaining and disruptive. Sustaining technology depends on improvements to already established technology over time. Disruptive technology, on the other hand, often lacks refinement, has performance issues in its early stages, appeals to a limited user base, and may not yet have any practical use (as was the case with Alexander Graham Bell's electrical speech machine, later marketed as the telephone). By contrast, the vacuum cleaner was an innovative device born out of a non-powered bristle cleaner still in use today. At the turn of the 19th century, electricity was the greatest technological invention and ushered in a sweeping line of new powered devices—from early automobiles[2] (yes, it's true) to trolleys, recording devices, and the telegraph. It wasn't a great leap to add an electrically powered bellows to a bristle cleaner. This is an example of a non-disruptive innovation.

Disruptive technology innovation, on the other hand, challenges the end user to rethink how they conduct business or complete processes. It creates entirely new markets, such as electric and telephone companies, online retail, and ride-sharing. That said,

---

[2] US Department of Energy, <u>https://www.energy.gov/timeline/timeline-history-electric-car</u>

disruptive tech can also involve new ways of thinking about doing processes, such as delivering goods, managing people, marketing, or banking.

Cryptocurrency is the most controversial area of tech today, and also the most misunderstood. Blockchain technology is not difficult to understand, but mastering the development skills to implement and secure it requires technical brilliance that few possess. I love using cryptocurrency as an example of disruptive technology because it's one of the most feared technologies ever developed. In the history of banking and finance, no idea has faced more opposition than the notion that crypto could replace fiat as the world's base currency. No one is more terrified than the U.S. Federal Reserve. To date, no strong legislation has been passed forbidding the trading of this technology. In fact, the attitude has been more of a 'if you can't beat them, join them' approach, as evidenced by the development of altcoins like U.S. Dollar Coins (USDC) and USD Tether (USDT). The reason for this is clear: blockchain is bigger than any single world government, so I can assure you, crypto is here to stay.[3] (Now, if we can just perfect quantum qutrit teleportation to create a truly non-hackable blockchain.)[4]

Another example of disruptive technology is a software platform that measures Objectives and Key Results (OKRs) to more efficiently manage projects, departments, and employees. Andy Grove, the creator of this business practice, taught it to John Doerr while he was an Account Executive at Intel in the mid-1970s.

---

[3] Cryptocurrency Space Needs Room To Grow: SEC Commissioner, Osato Avan-Nomayo, bitcoinist.com, August 05, 2019

[4] Daniel Garisto: *Qutrit Experiments Are a First in Quantum Teleportation, Scientific* American; August 6th, 2019

Doerr later introduced this management methodology to a startup called Google in 1999, while he was an early investing partner at venture capital giant Kleiner Perkins. Larry Page, CEO of Alphabet and co-founder of Google, is quoted as saying;

"OKRs have helped lead us to 10x growth many times over. They've helped make our crazily bold mission of organizing the world's information perhaps even achievable. They've kept me and the rest of the company on time and on track when it mattered the most." [5]

Since then, other disruptive companies like Uber, Twitter, and LinkedIn have adopted OKRs as part of their management methodology.

At the time of its conception, OKRs hadn't been proven to be any more effective than countless other methodologies. However, after decades of success among some of the world's largest tech giants, it's now considered the gold standard for startups because nothing to date has proven more effective at growing a company quickly and efficiently. One could argue that OKRs have been around too long to be considered disruptive, but that's not the case. It's still not a widely adopted business practice, and new Software as a Service (SaaS) platforms are continuing to fight for widespread adoption. Disruptive companies like Profit, Perdoo, and Weekdone are starting to catch fire as more businesses realize the benefits of this methodology. These companies face the challenge of converting late-stage early-majority and late-majority markets.

There is a subcategory of emerging technology with great potential to be disruptive: 'Frontier Technology'. These technologies are typically in deeper tech areas

---

[5] *Measure What Matters: How Google, Bono, and the Gates Foundation Rock the World with OKRs*, Doerr, John (2018). Penguin Publishing Group. p. 31

that are just emerging from R&D but are not yet at mass market commercial adoption. Today's sectors at the forefront of Frontier Tech include Space 2.0, machine learning, quantum computing, augmented reality, autonomous vehicles, and digital manufacturing. According to a report by Harvard's Kennedy School of Business, 75% of frontier startups fail to earn a positive return and go under within the first five years. Sixty percent of founder-CEOs are replaced by the time a startup raises Series D financing. The report goes on to say;

> "Frontier Tech entrepreneurship is often solution-led, driven by asking which potential applications for an invention should be commercialized first. This is in contrast to problem-led entrepreneurs who search for a solution to an unmet customer need."[6]

Sales executives for frontier tech companies have their work cut out for them, as the cards are stacked against them. Unproven tech, often fresh out of R&D, requires overcoming new obstacles before the market will accept the idea of radically changing how business is conducted. Therefore, frontier tech requires a sales methodology that is as forward-thinking as the technology being introduced.

If we've learned anything in this Information Age, it's that technology is always evolving and is sometimes replaced altogether by new innovations that may be disruptive —but not always. So how do we sell these radical and threatening products? The old ways of doing business aren't going to get the job done anymore. Strong relationships and an effective elevator pitch will only take you so far when asking a company to reconsider its supply chain methodology, adapt its culture to include a fresh approach to

---

[6] *Reimagining Investing in Frontier Technology*, Ash Carter, Laura Manley, Susan Winterberg: Harvard Kennedy School Belfer Center. June 12, 2019. https://www.belfercenter.org/publication/reimagining-investing-frontier-technology

management, or, even more drastically, overhaul its IT infrastructure so it can continue to do business at the blistering speed the industry demands.

When writing this book, I crafted its appeal for the seasoned salesperson who already has a solid understanding of sales basics—such as the steps of sales, overcoming objections, and delivering effective presentations. It is my hope that you will meld your unique style with the methodology I present, so that you become highly desirable to disruptive tech companies looking for a Rockstar Account Executive, Account Manager, Sales Engineer, Sales Director, or Revenue Officer who simply 'gets it'. If you love carrying the bag and are paid what you're worth to be the smartest person in the room, then this book will be your guide to winning disruptive tech deals in the most exciting time of technological innovation the world has ever known.

James L. Toland, Palo Alto, CA - 2024

CHAPTER 1- ENTER THE JEDI

When I walk into a boardroom to prepare for a presentation, I get into the mindset that everyone in the room will be very skeptical of the claims I'm about to make, regardless of what stage of the technology adoption life cycle they are in. That said, they desperately want me to convince them that, as strange as my product may seem at first glance, it's worth looking into. At some point, they will have to part with hundreds of thousands, if not millions, of dollars if I succeed in closing the deal, which is a hard pill to swallow for most C-suite executives. Because we're talking about game-changing technology designed to save or generate money for companies with a net worth of over $100 million, it's not going to come cheap. This is the first hurdle to overcome. The initial meeting should never be about price—it should be about the technology. Whatever the product is, there's some pretty nifty tech behind it, and that's what you need to know inside and out.

Sales luminaries have expounded on the importance of finding a customer's 'pain points'. But we're not selling minivans here. When pitching disruptive tech, most customers don't even know they have pain points. They may, however, have 'business problems' that already have an effective technology solution in place. As far as they're concerned, it's the best technology available—until you call and tell them otherwise. Disruptive technology comes to market to solve these business problems more effectively. The difference between pain points and business problems is subtle but important. Think of a business problem like a math problem. A CFO doesn't experience a pain point when balancing the books—that's just part of the job. However, the business problem that needs solving is how to monetize a new platform. That's not a pain point; it's just one of many challenges businesses face. A pain point is when the CFO can't get accurate sales data due to a technology failure, human error, or a lack of effective standard operating procedures (SOP).

Entrepreneurship, since the beginning of industry, was born out of the need to solve both kinds of challenges. A business problem might be getting a harvest to market and still making a profit, while the pain point would be not achieving the desired profit margin. The wheel was invented because carrying the harvest on one's back caused pain. If you're selling disruptive technology, you've recognized that taking a harvest to market with a hired trucking company is not as profitable as doing it with a fleet of autonomous semi-trailers. The farmer was perfectly content solving their business problem with a hired trucking company offering the lowest rates in the industry. The cost of doing business isn't a pain point until there's a more efficient or cost-effective method available.

There are five levels of the Jedi Order, with Grand Master being the highest rank. As far as I know, in the Star Wars universe, Yoda is the only one with this honor. You don't necessarily need to be that smart about your technology—that honor is reserved for the engineers who designed it. Luke Skywalker is a Master, and Anakin, his father, is a Knight. Be like the Skywalkers. You are not a Padawan or a Youngling. You are a master of the technology, and you're confident that your product will revolutionize your customer's processes and directly impact their bottom line like no other technology to date. That is the mindset—the force—you need to believe in with every fiber of your being. It's okay to be confident. Use strong words in your presentation, and avoid ambivalent words like *might*, *possibly,* and *maybe*. Leave no room for error. Testing will inevitably be part of the sales process, so don't use it as a roadblock. Hopefully, you're not bringing a beta version of your tech to the customer, so make it clear that your two companies will work closely together to ensure a successful onboarding. Leave it at that. It's going to be a long road of compliance certifications, pilot programs, and collaboration between engineering departments, so stick with what you know—your product rocks. If they don't agree to try and make it work for their company, they need to explain why out loud to you.

In the old days, we had Sales Engineers (SEs) to help with the heavy lifting regarding features, implementation, and integration. If you're heading the account and it's your pitch, you need to be the smartest person in the room, even if you're lucky enough to have an SE on the project. At the end of the day, the SE will help you immensely throughout the sales process, but it's you who will ultimately close the deal. CompSci degrees, MBAs, and GitHub influencer scores matter little if you don't come across as the unchallenged expert in the room. Establishing trust with the customer is paramount if you want to move forward. No pressure, right? I always say the first pitch is the hardest, but I've outlined below exactly how to do it.

Be a prepared Eagle Scout. No matter how well you know your product, make sure you've practiced your deck and that it has pop and sizzle. It will help you come across as confident and secure in your role. You can't wing it with boring slides and consultative tactics like, "*I'll look into that*" or "*Great question, let me find out for you.*" The deck must be bold and rock solid in its goal to present the product in the best light possible. Four criteria need to be adhered to during the first presentation:

1. Is it clear what the product is within the first 2 minutes?

2. Is it clear what the technology behind the product is?

3. Did you challenge the customer's current technology or methodology or provide a better solution to a known business problem?

4. Can the customer see the future as clearly as you do?

Before you walk into the pitch meeting, you should already have gathered all the necessary information about the customer's needs—through phone calls, emails, virtual meetings, or in-person discussions. That's important groundwork. You have one shot at this, so don't waste it gathering information you should already have.

Once you've dazzled your audience with what the product is, spend the majority of the presentation going into the technology. Assume that everyone in the room is smart enough to handle a technical discussion, and don't hold back. If they can't keep up, then you've got the wrong audience. Next time, make sure the room has the expertise necessary to influence the outcome. You don't always need the decision maker in the room, but if the influencers can't grasp what you're presenting, how can they pass the information forward? The inevitable outcome of dazzling the room with your understanding of the tech is that you'll establish yourself as the smartest person in the room on this particular subject. That last part is important. If you're pitching the CTO, CIO, or CISO, you better stick to technology. If you're pitching the CFO, you better know your numbers cold, and if the CMO is glaring at you over their iPad, you better have a plan that makes them look like a marketing genius. Know your audience, but remember—the technology will be what pushes the deal forward.

Now comes the fun part: challenging your audience to consider that the way they're currently doing a particular process is no longer the best way. Whether your product relates to delivering goods to customers, supply chain vendor management, or real-time sales analytics, position it as the best of breed today and for the foreseeable future. It's critical that you support this bold claim with as much data as possible. Every executive wants the best for their business, and that means they listen carefully to thought leaders who inspire

them to consider new ways to solve business problems. That's why you got the meeting in the first place. More on thought leadership and evangelism in Chapter 9.

I have a slide I love that simply asks, **WHAT IF**? Every innovation starts with that question, and it's a great way to kick off a conversation about turning a company's processes upside down. It's a direct enough question that, if you're lucky, you'll get an immediate answer. If the answer is something like, *"That's interesting. I'd like to know more about that,"* you've just leveled up. If the answer is, *"No, I'm not interested at this time,"* consider yourself lucky—you won't waste the next 6 to 12 months on a project only for it to die in executive committee. Fail quickly and move on with confidence. Call that company's biggest competitor as soon as you get to your car. If you successfully take down the competitor, you can bet the naysaying executive will come back to the table as soon as they realize they're getting their butt kicked. Don't be afraid to sell full life cycle solutions with long sales cycles. If you do your work and stay on task, you'll close the deal eventually. It might take a smaller pilot program at first to prove the concept, but have faith in your product—even if it's not fully polished. If you're truly selling a game-changing, disruptive technology, your customers will want to develop with you.

If you're working with an emerging technology, you're likely part of a brand-new startup, so you'll have little-to-no brand awareness or an established track record. This is probably the most challenging position to be in because it's hard to ask a client to take a chance on your product when there's a risk you might go under. If they've already been burned by non-starting startups, you'll have an even tougher time overcoming this objection unless your company has solid financials and support from well-known VC firms. Keep this in mind when considering your next position at a brand-new startup.

Here are a few things to keep in mind when vetting your next opportunity:

- Are you significantly vested, so it's worth taking the risk of failure?

- Does the company have a long enough burn-down or runway to get its product to market?

- What is the marketing plan to gain traction, and will it be enough?

- How credible are the founders and the VC backing the company?

- Most important: Is the product fully baked?

It's okay if you're starting small with plans to expand your product line. Think Alphabet before Google. As long as the product has credibility that you can speak to when overcoming objections, you have a good chance of success. In the case of SaaS solutions, it's best to limit your marketing to early adopters who can get the platform for free or at a low cost. Their user feedback is critical, and you need to grow your user base quickly so you can establish credibility. If you go after strategic accounts early, you'll get your teeth kicked in when they realize your product isn't ready for prime time.

It's my belief that software companies should limit their sales team to inside Sales Development Representatives (SDRs) or Business Development Representatives (BDRs) in the first two years after launch. Outside strategic Account Executives (AEs) are expensive and temperamental, and they won't be effective going after strategic F1000 customers if the product isn't solid. In short, make your mistakes early with smaller customers before hunting for whales.

Much has been written about identifying your best potential customers by categorizing them into groups: innovators, early adopters, early majority, late majority, and laggards. If your tech is less than five years old, you must stick exclusively to the first two categories. I can't stress this enough. The other three groups will come in time. If your company is entering its growth phase and you're targeting the other side of the chasm (the early and late majority adopters), your work is cut out for you as a salesperson. You'll face more skepticism, possibly more competition, and it will take longer to close deals. There will be more pressure to grow revenue quickly.

Identifying what stage a company is in requires knowing the tech they're currently using. It's like looking for smoke to find fire. You already know their tech is obsolete, and your tech will replace it. If they're early adopters of unrelated technology, they probably know their tech is outdated and are looking for the next best thing to gain a competitive edge. How do you find the smoke? That's the 64-bitcoin question and may be the subject of my next book. If you have a crack business development team, let them worry about that. Most likely, if you're getting the word out about your technology, early adopters will come to you—but that alone won't make quota.

Keep your ears to the ground by reading trade publications, paying attention to press releases, and tracking companies similar to your own that also have game-changing tech. They don't have to be competitors—synergistic companies can also offer insight. I once worked for a company that had been an Apple manufacturing partner for 25 years. Since their product was MFi certified, I worked closely with Apple Enterprise AEs and SEs, as well as mobile device management companies like Jamf and AirWatch, to learn of large Apple device deployments because our product enhanced Apple's enterprise capabilities. Five years ago, if I had told Honeywell, Motorola, or Zebra product managers that their Windows-powered mobile computers would be rendered obsolete by the iPhone, they would have laughed me out of the room. But in 2018, Microsoft announced it would no longer support its mobile operating system after 2020, leading to a surge in Android and iOS device sales. Android devices are cheap and have broad appeal, but Apple is winning logos faster than ever because they offer best-of-breed security, speed, and integration. Sure, the iPhone costs more, but it also does more. Apple's virtual scan engine is superior to any firmware scan engine, replacing not-so-smart barcode scanners and giving employees more computing power to handle complex tasks. This is the very definition of disruptive technology, as it has changed the way we manufacture, design supply chains, and train workforces.[7]

As I said earlier, the first pitch meeting—whether virtual or in person—is the hardest part of the sales process. Your goal is to convince your audience that your product will revolutionize their business and that you're the smartest person in the room on this subject. If you nail it, there are two possible outcomes: 1) it's not for them, or 2) it's interesting enough to explore further. If the latter happens, the real work begins—proving to the decision-makers that your technology actually delivers on its promises. However, if someone in your audience thinks they're the smartest person in the room when it comes to your technology, they've just identified themselves as a laggard in the technology adoption life cycle. There's little you can do to convince them otherwise, so it's best to walk away. They're probably still using a flip phone.

---

[7] Full disclosure I do not have any data to support the performance of iPhones over other brands. There are just too many use cases to compare devices head to head. I have been observing however a shift of large enterprise companies adopting Apple devices over Android in recent years and I have heard from Android based mobile device manufactures that they recognize Apple is gaining market share.

CHAPTER 2 - JUST ENOUGH TO BE DANGEROUS

I should probably make it clear that these chapters aren't necessarily illustrating steps in any particular order. I started with the first pitch meeting because it's the culmination of weeks of work and preparation. To be frank, it's just a more exciting place to start a book about selling game-changing technology. This chapter will focus on preparing for the first pitch meeting(s). Sometimes there's more than one. This is especially true if you're working with a company in the U.S. that has foreign headquarters. You might have to take the show on the road to Switzerland if you gain traction in the U.S., depending on how the company is structured. I've pitched in New York City, only to be told I needed to fly to Seattle and do it all over again, in front of the same team plus one straggler who apparently had to be involved for us to move forward. Practice makes perfect.

Now that we've established you're a Jedi Master in your technology, it's time to learn enough about your customer's business processes to be dangerous. Identifying the use case(s) or business logic of your customer allows you to speak their language and tailor your pitch to their specific needs. Have you ever pitched to a prospect only to find out they have no present use case or processes that require your technology? That can be an interesting predicament, especially if your tech requires compatibility, like iOS devices versus Android. But if your technology is so new and different that the customer hasn't even imagined it was possible—for example, same-day delivery of products to people's homes via a flying drone or autonomous vehicle—then you're in the best-case scenario.

Imagine pitching robotics pioneer Scott Hassan's product, 'The Beam' by GoBe Robots, to Mark Zuckerberg. It might go something like this:

"What if Facebook users could hire a robot anywhere in the world that they could control to explore exotic locations? And what if they could experience what the robot's sensors sent back through 3D goggles tethered to their smartphone with an enhanced virtual overlay?"

Then Facebook would ask,

"What would something like that cost?" and you'd say, *"It would cost whatever it takes for you to net $100 million."*

I can say that because I know what the product costs, and I know how Facebook could monetize it for a 200% profit per hour of use, based on market research my company commissioned.

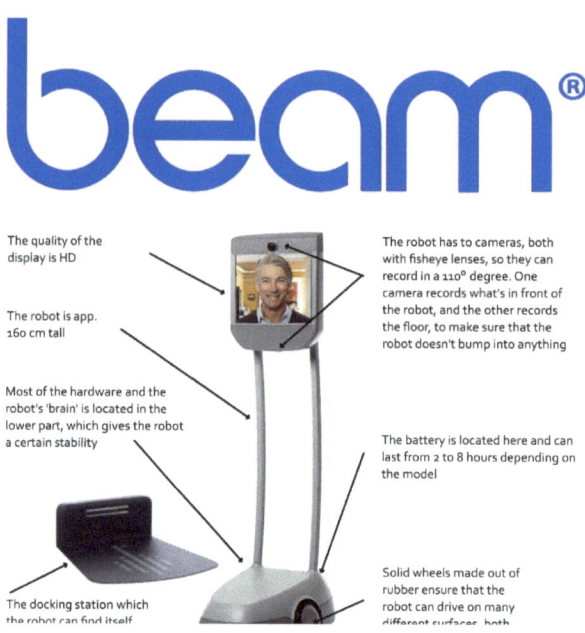

Knowing enough about a customer to be dangerous means you've collected your own data and analytics on their company's products, performance, market penetration, etc., and you're confident enough to speak about them when necessary. Moreover, you're an expert in your field, so you already know what typical choke points are in supply chains or inefficiencies in logistics. Speak to this, and you'll gain even more credibility and trust with your customer. I remember once going on a sales call with a channel partner, and he actually asked the CISO, "*So, what do you guys manufacture?*" Before the executive's jaw hit the floor, I quickly answered the question for my colleague, buying us a few more minutes to make our pitch. If I hadn't, I'm sure the Chief Security Officer would have been called away to an emergency before our pitch even started. As it turned out, I closed the deal with very little help from the channel rep.

To date, a plethora has been written on the subject of Challenger Sales methodology vs. Solution Selling. I'm not going to weigh in on the benefits or shortcomings of either—only to say it's important to have some sort of methodology, even if it's a hybrid. If you're not familiar with either, I highly recommend looking into it on your next business flight. Like any sales technique fad, there's good stuff to be had if you keep an open mind—and a few grains of salt handy. It's important to take your leadership's book recommendations seriously, or at least read the Blinkist summary, so you can weigh in at the next all-hands meeting.

If you're going to challenge a customer to think differently about how they do business, do it armed with enough data and analytics to support your claims. The same goes for providing solutions to known problems. That's really the fundamental difference between Solution Selling and Challenger Sales methodology: knowing what the business problem is versus showing them there's a problem. In the aforementioned example of Hassan's *Beam*, Facebook doesn't have a business problem with not being able to teleport users to Machu Picchu for a real-time look around. If they did, you'd have a solution for them. But we're talking about disruptive technology that was born from someone asking, "*What if we could?*"—for no other reason than it seemed like a good enough idea to build a company around.

Before I wrap up this chapter on intelligence gathering, I'd like to offer some personal advice: spend as much time researching and collecting data as the opportunity is worth. It's always a good idea to schedule a few calls to learn from the customer what interests them about your product and what their business goals are. After that, do your own research, collect data and analytics, and run it by your team to ensure you have a cohesive angle. Don't spend weeks researching a project worth $50k when you should be proactively landing larger accounts. I've known AEs who've been sacked because they spent too much time on smaller deals, or inversely, not enough time on a whale of a prospect. Another piece of advice is to create a sales plan targeting companies in the same vertical so you can use the industry-relevant deck and data over and over again. Your pitch will get better, and when you land a win, you can use it to build credibility with the next similar company.[8]

Knowing your customer's business and being able to speak to known business problems requires humility—you don't want to come off as though you know an executive's business better than they do. That's not the right kind of arrogance to win their hearts and minds. A better approach is to speak confidently about what you know about their industry, defer to their expertise when possible, and share your findings openly as if they already have the data.

Let them know that you know what they know,

and that they should be interested in a little secret

that you know, but they don't—yet.

---

[8] Make sure you are permitted to share your client's testimonial. Get it in writing if you can so you don't violate any NDAs enforce.

CHAPTER 3 - THE FUTURE IS HERE

How many times have you heard someone say, 'sell the dream'? I don't like this idiom. I'm sure it was coined by some adman on Madison Avenue in the `50s. Hopefully, we're not selling dreams anymore, like in the `90s when there seemed to be more vaporware on the market than software that actually worked. If you were around then, you might remember IBM's Ovation or Xanadu, to name a couple of epic fails. Selling the future is what this book is really all about. By taking a position of confidence—although the future may be uncertain—the data strongly shows that your company checks all the boxes of a stable and viable frontier technology leader.

Will Poindexter of Bain & Company wrote in the MIT Sloan Management Review that CIOs are, for the most part, reluctant to make technology investments for fear of failure. He suggests asking these simple questions when evaluating new technology adoption:

1. What business or customer problem are we trying to solve?

2. Is the investment worthwhile?

3. How can we test new ideas?

4. How do we measure results?

If you address these questions right from the start, you'll be speaking the CIO/CTO's language and quickly building trust with your decision-makers. As I listed in Chapter 1, there are four critical objectives to a successful deck:

1. Is it clear what the product is within the first two minutes?

2. Is it clear what the technology behind the product is?

3. Did you challenge the customer's current technology or methodology or provide a solution to a known business problem?

4. Can the customer see the future as clearly as you do?

'Can the customer see the future as clearly as you do?' is where you detail all the steps in the project, including testing and measuring results. Also, are the next steps of the project clear, and is the objective in sight? 'Can the customer see the future?' also means: Can the customer imagine what their world would be like if they adopted your technology? I know that sounds like selling the dream, but it's not—because it's a real and viable option, dependent on working technology. What the customer should be imagining, with a Cheshire Cat grin, is how the product will save them operational money while increasing revenue. That's what the future needs to look like by the end of the meeting. You'll still have to prove the concept and go through all the testing and pilot programs, but throughout the project's life cycle, the decision-makers and influencers can never lose sight of what the future looks like. Going forward, that is your penultimate responsibility —after keeping the train on the tracks. Going back to the Beam example, keep reminding Zuckerberg what the financial windfall will be when 3 million users spend $2 every day to roll around the Taj Mahal: That's $2.19 billion, Mark.

Whatever future you and your customer agree on needs to be documented in the form of a Scope of Work (SoW). This will serve as the outline you stick to and, if necessary, amend as the project progresses. The SoW is only needed if you intend to test and/or integrate the tech into the customer's existing systems. It's also essential for any custom development projects. I find it very effective to use a shared project portfolio management (PPM) platform to keep the SoW on track and the future always within grasp. (Sometimes referred to as a Project Management Platform.) There are plenty of PPMs to choose from, but ask your customer what they currently use and get on board with it. If you've never used a PPM like Hive, Jira, Basecamp, or Microsoft Project, I strongly recommend getting up to speed. They all have similar features that keep everyone on track and in communication. I also like to use the internal communication features within PPMs to stay in touch, as it gets quicker results than email and fosters a more familiar exchange, which hopefully leads to a closer working relationship.

The next chapter will touch on additional methods of digging in and getting *sticky* with your points of contact. PPMs allow you to schedule milestones, allocate resources, create timeline visualizations, track time, manage tasks, collaborate, and use Slack-like internal messaging. It's safe to say that full sales cycles often fail due to unclear next steps, miscommunications, lack of interest, or lack of resources caused by poor planning. Utilizing a PPM will prevent this from happening and most likely lead to a quicker close. Now, who doesn't want that?

If Project Managers (PMs) are thrown into the mix, as they often are, you can bet they have a PPM for agile project management in place, and YOU will become their favorite salesperson if you adopt their platform. If you're working with Product Managers on the project, they may also have their own platform that could work, but learn their language first—because you don't want to show up at a s*crum* without a developer as backup unless you can write code with the best of them.

Project management best practices are complex and nuanced and are responsible for some of tech's biggest players, such as NetSuite, Oracle, and SAP. Business logic is at the heart of any project management solution. What logic are you going to use to manage your deals? If you're not sure, spend some time writing out all the steps you foresee using in your sales plan, and consider what results or behaviors you anticipate encountering. You're essentially going to create a workflow—much like developers do when programming a database interface. From this workflow, patterns will form, and benchmarks will be revealed. If you plug all this data into your CRM or PPM, you're essentially creating an algorithm that will guide your business logic moving forward.

If I haven't taken the decision-makers and influencers out to lunch within the first 30 days of working together, the project is probably a nonstarter. Again, this is a major difference from widget sales, where deals are often started and won over prime steak dinners. Schmoozing clients as part of the pitch wouldn't fly with emerging high-tech deals. However, it does help bring the teams together after there's been a meeting of the minds and a clear SoW in place. Building relationships with your points of contact can never hurt—after you've committed to working together for 3-6 months. Good old-fashioned relationship building comes after the trust-building mechanisms. That's the fundamental difference and the impetus for writing this book. It's frustrating to watch colleagues struggle in the field when they've come from other sales environments with *old school* methodologies. I once worked for an IT services company with a trunk full of really forward-thinking strategies. They thought it was a good idea to hire biz dev reps (BDRs) who had previously worked for Yelp. The sales methodology taught at Yelp's boiler room was in no way compatible with the nuanced consultative technical sales necessary to sell IT optimization. The sales leadership believed that because these reps were accustomed to making 60 calls a day, they'd be effective in persuading key decision-makers to meet with an AE. I've never seen a more frustrated team of talented salespeople get their teeth kicked in day after day, resulting in record-high turnover. The ones who eventually succeeded learned they needed to slow down, ask a lot of questions, and position themselves as trusted advisors rather than telemarketers. Obviously, this required brushing up on the technology and taking an interest in the industry they were targeting.

Harkening back to the Star Wars analogy, they were Padawan-level Jedi on the tech, with just enough understanding of smart manufacturing to be dangerous on their

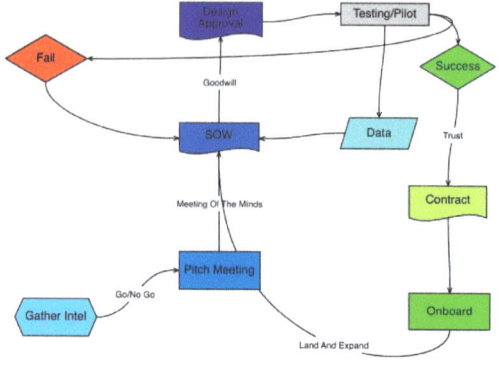

Figure 1.   Figure 2.

best days. If this is where you are in your career, heed my advice: the methodology in this guide will help you get considered for an AE position faster than you might think possible. The difference between making $65-90K as a BDR or AM and making $300-600K as an AE comes down to this: what Jedi level are you about your product, and how well can you manage a full-cycle sales project? **Spoiler alert**: That's really all there is to it.

Figure 1 above shows the steps of a typical emerging technology sales cycle. Some of the links have mechanisms. These triggers are essential and often overlooked components of the sales machine. Figure 2 shows a typical software development cycle. The flowchart starts with gathering intel.

    As you can see, the Software Development Cycle is similar to my sales model, as the two are closely linked. It goes without saying that trust-building practices should be exercised throughout the sales process, especially after the SoW has been approved and before you get into the pilot phase.

The takeaway from this short chapter is: stay organized and keep the promise of the future alive. Moving the project along at a steady pace is essential, and you'll gain goodwill credits when the project encounters hiccups and seems like it might be going south. You can always fall back on your goals and allocate additional resources to get the job done if necessary. The scope of the project, along with the objectives and goals, are all outlined and agreed upon by the customer in the SoW document, so use it as leverage to stay on track. Once you've reached a 'meeting of the minds', you're now in a partnership that depends on both parties' willingness to succeed. This is a major difference between selling emerging technology with few rivals and well-established tech that requires a more aggressive strategy—outperforming competitors and cutting deals based on price and value. The stakes are much higher with emerging disruptive tech because there are more unknowns and greater risks involved. It's essential that you form a strong bond of trust and foster a collaborative spirit. It's going to get weird, I promise you. Tests will fail. Expectations will be dashed. Confidence in the outcome will waver on both sides. Adopting the mentality that great innovation comes with great risk will bond your teams and help you weather the storm together. In the immortal words of the great American author Hunter S. Thompson:

"When the going gets weird, the weird turn pro." [9]

---

[9] *Fear and Loathing In Las Vegas*, Hunter S. Thompson; Random House, 1971

CHAPTER 4 - INTELLIGENT EFFORT

The great art critic John Ruskin once wrote, "*Quality is never an accident; it is always the result of intelligent effort.*" This quote can often be found on sticky notes on developers' monitors or scrawled on startups' vision boards. The wisdom is often associated with software development because so much testing is required to create a quality product. There's also a joke often found written on bathroom stalls at tech companies that simply states: *FAST, GOOD, or CHEAP – PICK TWO.* It's a sad reality, but one that has been the challenge for many portfolio companies trying to build the next great SaaS solution or consumer app. It takes great intelligent effort to make a product work, and you, as the sales executive and sometimes project manager, need to be an active part of this—the trial stage.

The first 30 days after a meeting of the minds, I call the *trial stage* because it's when the customer tests your product in or on a small, controlled group, inevitably trying to see if they can break it. They haven't spent any money yet, and they're still dreaming of lollipops and ice cream cones. Anything and everything could go wrong, and you'll never know it if you aren't staying in front of the champions—contacts who are hopefully pulling for the success of your product within the target organization.

I like to install myself in my client's field of vision. Make the best use of your visits to get to know the cast of characters involved in the project. You won't have time later when the tires hit the road, and you're going 120 mph toward a speed bump. Building goodwill early will help you in the long run. I can't emphasize this enough, even though I think I already have. It's a key sales mechanism, like a Scope of Work (SoW) or a development agreement. Stacking up these mechanisms is part of the strategy to keep the project moving forward. Every stage of the project has a trigger that starts the next stage. You NEVER want to find yourself in a pit of data darkness.

Data Darkness is simply anytime you do not have a flow of intel regarding any process—in this case, the sales process. I've experienced how frustrating this is, especially when we have so many wonderful technological tools to dashboard these data sets. Lincoln Murphy, in his book <u>Customer Success: How Innovative Companies Are Reducing Churn and Growing Recurring Revenue</u> [10], speculates that about 25% of opt-in free trials convert to paid subscribers. He also states that 25% is a goal, not a measure of success. If you're at 5%, strive for 7%, then 10%, and so on until you get as close to 25% as possible.

---

[10] Marian Martinez (Author), Nick Mehta (Author), Dan Steinman (Author) *Customer Success: How Innovative Companies Are Reducing Churn and Growing Recurring Revenue*; Wiley publishing February 29, 2016.

In my experience, limiting data darkness will greatly improve your opt-in free trial conversions. Think of data darkness as the dark side in George Lucas's Star Wars universe—the epic struggle between Jedi-backed rebels and fallen Jedi knights who embrace the Dark Side of the Force. I know it's corny to reference, but the common denominator is just this obvious. Taking it one step further, I suggest seeking out the balance points of an opportunity so you can quickly recognize which deals are likely to move to close. Speculating is, unfortunately, part of finding the right balance between working on 'late-stage deals moving to close' and 'struggling projects growing colder.' It's never easy, but the alternative is far worse: misappropriating resources that result in low conversion rates. Adopting an OKR (Objectives and Key Results) management accountability methodology goes a long way toward easing this pain.

Much can be learned about sales from the hospitality industry. I've always said I don't trust anyone who hasn't worked at least one day in the restaurant industry. I've read that the practical definition of hospitality is 'anticipating a guest's needs before they do.' If you apply that simple notion to your sales process, you may find it easier to win the customer's heart and mind. I'm passionate about fine cuisine, and one of my favorite restaurants is Eleven Madison Park in New York City. Its former owner, Danny Meyer, CEO of Union Square Hospitality Group, wrote a fantastic book called <u>Setting the Table: The Transforming Power of Hospitality in Business</u>. Meyer examines the power of hospitality not only in the food service industry but in business at large. Hospitality is a powerful act of kindness that requires dignity and grace.

He states:

"We are in a very new business era. I'm convinced that this is now a hospitality economy, no longer the service era. If you simply have a superior product or deliver on your promises, that's not enough to distinguish your business. There will always be someone else who can do it or make it as well as you. It's how you make your customers feel while using your products that distinguishes you." [11]

The most wonderful aspect of working for a visionary tech company is that you don't have to hard sell your product or engage in sales trickery to win a deal. In fact, those practices will most certainly not be tolerated by anyone willing to take a great risk with you. Perhaps that's why I find my career so exhilarating and fulfilling.

I've found that the simple act of exchanging information can strengthen relationships and build goodwill quickly. Sending pertinent articles, third-party studies, and performance data throughout the sales cycle as part of the ongoing conversation is typically well-received. I also like to ask a lot of questions to better understand the customer's point of view—particularly the company's priorities and how our project fits into their master plan. This kind of intel is best extracted in a casual setting rather than addressed as a bullet point in an email or deck. I'm never afraid to ask hard questions like, "*What sort of challenges do you personally face when trying to update your essential technologies?*" I've received answers ranging from 'a shortsighted boss' to 'a penny-pinching CFO.' My favorite answer was, "*I love what you guys have developed, and if it were my company, I'd have it running tomorrow, but XYZ company (a competitor) owns 20% of our company, so we need to go with them. Sorry.*"

---

[11] *Setting the Table: Transforming Power of Hospitality in Business*, Danny Meyer, Harper Collins, 2001

During the testing stage, Sales Executives must support the early user experience (UX). The world of UX design is honestly outside my comfort zone to discuss in-depth as a business problem, but I do know that products live and die by how easily—and, dare I say, 'effortlessly'—users are guided through an often-steep learning curve. Depending on the kind of field support and professional services you offer, this heavy lifting may be assisted by an SE or technical team. Be transparent if you're not comfortable answering technical questions about security protocols or integration with third-party software. Whatever you do, don't try to bluff your way through tough questions just to appear as the smartest person in the room. Even if I don't need a Sales Engineer on a call, I like to have one there just in case I get a sticky question during the testing stage. It's better to have one and not need them than to need one and not have them.

Establishing yourself as an engaged thought leader who adds value will go a long way toward strengthening the foundation on which your deal is built. Have fun with your clients. Be confident and transparent while also being mindful and compassionate. Smile often. Tell the truth and be patient. It's going to be a fun project.

CHAPTER 5 - THE PARADOX OF DISRUPTIVE TECHNOLOGY

The age or size of the company interested in your product does not necessarily define what stage of new technology adoption they're in. Early majority adopters could be a Fortune 100 company or a well-funded startup. It really comes down to their leadership's vision at the present time. That's why it's important to pay attention to industry news and follow the companies you have your sights on. While gathering intelligence about your prospect, pay special attention to recent technology adoptions and internal innovations that could give you an idea of where their leadership stands when it comes to early vs. late adoption of new technology. If your prospect is a startup but well-funded enough to afford your product(s), they are absolutely working with at least one venture capital firm (VC), which will almost certainly have its own management methodology already ingrained in the young company. This is a great opportunity for you if your technology supports their mandated methodology or vision.

Investing in startups is inherently a risky venture, so VC firms mitigate the risk by looking for characteristics in the candidate that will help them become successful. Besides the obvious appeal of the product, they assess leadership credentials, management methodologies, fiscal platforms, go-to-market strategy, revenue plans, and exit strategies when calculating risk. The opportunity for you, as an evangelist of your game-changing technology, is to appeal to both the VC firms and startups by showing that having your technology in the mix will ultimately help them mitigate risk.

When you say it out loud, it may seem contradictory to claim that you're an emerging technology with the potential to be very disruptive, while also mitigating risk. Nothing about new technology speaks to a long history of success. So, how do we ask a prospect to take a risk with early adoption of an unproven technology, when doing so could hurt their chances of getting additional funding? **This is The Paradox of Disruptive Technology.**

The answer is in plain sight. Behind every well-funded startup is a successful VC firm. The credibility of the firm backing your technology is your product's most valuable endorsement. Use it early and use it often. It should be front and center in your first deck. If your company is privately funded by its own leadership, then you have the strength of the founders to highlight. Companies like Sequoia, Accel, Kleiner Perkins, and Founders Fund have numerous very smart partners always on the lookout for the next unicorn in tech. If you're looking for fire, start with the smoke the VC partners behind your tech are making. If you can convince your VC partners to promote your technology to their own portfolio, securing deals could be as easy as shooting fish in a barrel.

Pitch to companies in your pocket VC firm's seed portfolio by emphasizing that your technology will help them establish (not guarantee) credibility when they return for early round funding. Sometimes VC firms will insist on certain platform adoptions, such as in the case of Google's piloting of the OKR management methodology suggested by early Alphabet (Google) investor and partner John Doerr at Kleiner Perkins. Imagine if it was your tech platform that Doerr insisted be in place if he was going to invest any more money. This is precisely how VCs assert themselves to feel secure about their investments.

When I'm considering joining a company, I always ask the headhunter which VC firms and, specifically, which partners are backing the company—because that's where I start prospecting. To that point, expand your network to include VC partners and try to make some friends on 'the inside.' After a few years of successfully selling to top emerging tech logos, your name and reputation for being a really smart person will precede you. That will not only help you get meetings with the industry's top movers and shakers but also make you a valuable employee and a potential candidate for an executive position if that is your goal.

Prospecting effectively is not something that can be easily taught, nor is there any one methodology that works better than another. There are too many resources and angles to consider that are industry-specific. With that said, I recommend working from the inside out, as I illustrated above. This is a radical idea, and you'll need to run it by your sales leadership first, as companies are typically protective of their investors, and it is highly unusual for an EA to cold-call them with a sales strategy. Make the case for why you feel it's important to address **The Paradox of Disruptive Technology**. If you get pushback, you can still forge ahead prospecting within your own VC's portfolio on the strength of your exclusive membership. You might also find that other VC firms will take notice of your approach, especially if the players backing you are high-profile.

I've heard it said in Hollywood that 'originality is a commodity.' One could say the same for Silicon Valley. When a company like Sequoia invests, many smaller, aggressive firms start campaigning as to why their money is just as smart. To that point, **DO NOT** presume to know the details of any investment agreements between parties. If, somehow, you do know, keep it to yourself—for the love of Steve Jobs' ghost. Nobody likes a gossip, and the old adage is very true: tech is a small world, and word travels at 1.5 Mbps. Don't be the person who upsets the apple cart to get the lone pear. Be cool and let your technology speak for itself.

CHAPTER 6 - PROOF IS IN THE PUDDING

Here we are on the eve of testing your technology in the field as a trial account. The SoW has gone through some necessary revisions, and a Design Concept (DC) has been presented for approval. This document is the fruit of hundreds of hours of development meetings involving every department, scrums, low, mid, and executive meetings, as well as board approval. It's a thing of beauty, and it's exactly what the customer wants. Concessions on both sides of the DevOps teams have been made, so it might not be all of the things the vision board initially recorded, but it's a great start to a long development partnership.

The project has had its fair share of hiccups and setbacks, but if you've used your project management platform successfully, expectations on both sides have been managed effectively. Hopefully, everyone is still on good terms, and your product management team hasn't put your headshot on their office dartboard. A long time ago, I figured out how to **NOT** push Product Managers' buttons, and it has served me well over the decades. Product Managers can be a vital part of the sales process, but you need to understand how they operate and avoid getting on their bad side.

A great analogy for this dynamic is the restaurant industry. There are essentially two departments that work closely together to bring you your meal: the front of the house and the back of the house, better known as the waitstaff and the kitchen staff. Both departments have a tiered management structure that works hard to keep their respective employees from killing each other. This is, of course, hyperbole, but not far from the truth. The kitchen staff is fiercely proud—they take great pride in creating and presenting their products in the best light possible. The waitstaff, on the other hand, is focused on the customer. Waiters make a living from tips, and since hospitality is about preemptively anticipating the guest's needs, waiters will do whatever it takes to make the guest comfortable—even if it means frustrating the hard-working kitchen staff with what engineers refer to as 'design creep.'

Can you imagine the reaction of a chef if someone ordered *blanquette de veau* [12] with the sauce on the side? I can tell you from experience, there will be bloodshed. Sales departments are notorious for promising design features at the expense of the product department. When this happens, it diverts the product department's much-needed resources away from other projects. This is why we have a SoW in the first place and a project management plan that includes an agreed amount of resources. If you ask for more than you budgeted, there will be fierce pushback from the Project Management

---

[12] A ragout of non-browned veal in a cream sauce with mirepoix (carrots, celery and onion.)

department, which may seem combative or non-cooperative. But that's not the case, so don't take it personally.

Imagine if a Project Manager walked into your office and took all your Mac power cords, saying their department's part of the project was more important than yours. It may seem like a reasonable request to add a bell or whistle to a product, but even the smallest engineering modifications can take weeks of man-hours. Be mindful of your last-minute demands and always consult with your Project Management team before EVER giving your customer the impression that their 11th-hour request is possible. There's a time and a place for revisiting the SoW and reworking the build, but this typically comes after testing and pilot data have been gathered.

DevOps is one of the fastest-growing sectors of technology as of July 2019 [13] [14]. The reason for this is simple: companies need development faster than ever to compete. Outsourcing coding and purchasing pre-fab blocks of code, as well as using launch automation platforms, are in great demand as software companies realize that new builds need to be rapidly deployed in days rather than months. Speed and accuracy are the names of the game, and forward-thinking disruptive companies like Morpheus Data, Refactr, and Oakland's LaunchDarkly are seeing record valuation growth. VCs are all in on DevOps companies this year, sparking a whole new job title boom for coders looking to cash in on their talents. The average salary of a DevOps engineer, according to Glassdoor, is $115,666.

---

[13] hackernoon.com, 8 *DevOps Trends to Be Aware of in 2019*, November 19th, 2018

[14] Grandview Research, *DevOps Market Size Worth $12.85 Billion by 2025* | CAGR: 18.60%, March 2018

When testing your technology in the field as a pilot rollout, be very careful to manage the end user's and decision maker's expectations. Early adoption of any technology comes with its own set of headaches, and for the end users, it often means more work outside of their comfort zone. Training and supporting the end user is paramount for ensuring a positive experience. If you're the AE, lucky for you, this responsibility falls squarely on the shoulders of Professional Services (PS). Trust your PS team and offer support whenever possible, but be mindful that you're on their turf now.

By the time a project is in pilot mode, you should already have a good rapport with your inside champions. You can step back and work on other projects that need your attention. Keep an eye on the progress, but don't react too abruptly when things get weird. Let your PS team handle it—they know how to keep the user experience positive. Everything I've ever learned of importance when it comes to managing a technology rollout has come from PS teams with way more field experience than I have. You don't become a PS without years of product management and sales experience, so when it's time for them to shine, take a step back from being the smartest person in the room. That honor now belongs to Professional Services.

By now, I hope I've illustrated ad nauseam why it's so important to have documentation in place detailing the entire scope of the project, complete with plenty of resource allocation on both sides of development. Emerging disruptive tech is rarely ready for primetime right out of the box. It's not uncommon to release free betas to gather user experience data and ensure platform compatibility early on in development. That's why it's called 'emerging technology', and it's why we must adapt our sales plans to effectively manage these unique projects.

CHAPTER 7 - MARCH ON AND FEAR NOT THE THORNS

Khalil Gibran said it best in <u>The Prophet:</u> "*March on. Do not tarry. To go forward is to move toward perfection. March on and fear not the thorns or the sharp stones on life's path.*" Let's assume the pilot program was moderately successful, but the customer is having a hard time justifying the cost of the product as it stands today. The deal is cooling off, and there have been some significant developments since the project started six months ago. You've done an exemplary job managing the customer's expectations and keeping the project moving forward. The customer respects and trusts you, but they've hinted that they don't have the budget to put the technology in play this quarter.

What went wrong in this scenario? The answer is absolutely nothing. This is just how disruptive tech deals go before they are won. Either the customer is afraid to change their technology because of the potential impact on their operations (lack of vision), or they truly don't have budget approval. If you did your job well and your technology lived up to the hype, it's most likely the latter. This part of the sales process will feel familiar—it's about overcoming objections.

It's negotiation time, and the customer is simply trying to get the best deal possible. Be confident that this is the case, and start asking questions like, *"What do you need to move forward to close this deal?"* If they're haggling for a better price, give them what they want—within reason. The best scenario you can have is your technology in play. Trust your product. Remember, your product is disruptive. It's going to change the way they do business or create a whole new business practice. You might have to wait for the world to catch up with you. Remind your customer of this and point to industry data that supports the radical future you've been promising.

Let's use the old saying, 'the quick and the dead,' to illustrate this hard fact. Point to examples like Napster and MySpace, or Yahoo and AOL. They all have one thing in common: they refused to adapt to a quickly changing landscape and rode obsolete tech straight to the grave. One of my favorite examples of disruptive technology triumphing over traditional business practices is the sad story of American Apparel. They went through two bankruptcies and a fire sale before they got on board with innovative retailers like Adidas, Google, Northern Grade, and StoreEnvy.

While American Apparel was opening some 200 traditional stores across the country—dealing with staggering overheads and staffing challenges—forward-thinking retailers were embracing technology platforms from Samsung and Microsoft, as well as startups like ZigZag Global and Pensa Systems, to enrich the shopping experience. They also solved a new retail business problem that didn't even have a name until recently: 'data darkness'—the period of time between a shopper entering a store and purchasing a product. Now, if you enter a store, your smartphone comes to life and starts recording your behavior. If you're a retailer and not subscribing to these analytics from Google or third-party platforms, you've missed the boat entirely. Even McDonald's is using this data and providing an automated customer checkout experience.

When athletic giant Adidas started opening pop-up stores offering rare and collectible versions of their apparel and shoes, they achieved sales numbers that far exceeded what any of their brick-and-mortar stores could do on their best day. By cutting customer acquisition costs in half in some cases, pop-up retailers have accepted the challenge thrown down by retail giant Amazon. Unless you are Walmart, whose core demographic actually enjoys walking the store in their Sunday best, forward-thinking retailers are adopting disruptive technology with a ferocious appetite.

If your customer is still pushing back, make sure they articulate exactly why. It's vital that they say it out loud for the record. It might just be an internal issue like poor quarterly numbers or leadership changes. If a company valued at $100 million says they don't have the money, they're not saying they don't have the money—they're saying they can't get approval, yet, to spend that kind of money. Be understanding of this development, and stick to your guns. I'm sure if you go back to your own leadership, you'll come up with a deal that saves the project from stalling, even if it means waiving the first three months' subscription.

CEO	Chief Executive Officer	all projects in tangent with department head
CFO	Chief Financial Officer	financial and revenue platforms
CISO	Chief Information Security Officer	security platforms and solutions
CTO	Chief Technology Officer	hardware and ERP
CIO	Chief Information Officer	ERP, CRM, IT, SaaS, data & analytics
CSCO	Chief Supply Chain Off.	supply chain and equipment
CRO	Chief Revenue Officer	sales and marketing solutions
CMO	Chief Marketing Off.	marketing, data and analytics
COO	Chief Operating Officer	supply chain, operations, equipment, tools, ERP
Head of Manufacturing		Hardware, equipment, tools, MFG-platforms, robotics

Whatever you do, don't lose hope, but be mindful of the delicate situation you've put your customer in. Yes, it's all your fault. If you hadn't shown them the future and successfully managed this project through the entire sales cycle, they wouldn't be battling with their board to get the funds needed to push forward into the relative unknown. It's actually a pretty great view of a company's evolution, so enjoy it.

The strength of my guide for selling disruptive technology is not in the rigidity of the plan but in its flexibility. However you need to achieve success, do it in partnership with your customer. From the first pitch meeting, drive home the notion that you are in this together, and at the end of the project, both companies will be better for it—running, not walking, side by side into the exciting future.

CHAPTER 8 - YOU'RE GONNA NEED A BIGGER BOAT

When Chief Brody is throwing chum and Jaws breaches for the first time, it becomes clear to him that this shark is bigger than he could have possibly imagined. What's so interesting about this scene is that Quinn, who is an apex predator on shore, knows he has to become a skilled hunter to defeat the true apex predator of the sea—the Great White.

Up until now, you've been pretty reasonable and well-mannered. You've built trust and goodwill with your customer. You've successfully onboarded your technology, even if it was just a pilot program. So, what's left to do? Plenty! Now the blood is in the water, and your predatory instincts should be kicking in. It's time to 'land and expand.' Early success with a partial rollout MUST lead to more user onboarding. If the original vision was to have everyone in the company use your product, that's the prize. If piloting in one department means expansion into another, that's your new goal. If only one plant is using your tech, why aren't the 25 other plants worldwide on board? I've made entire year revenue goals by going deep and wide inside existing customers. Typically, it's closer to 70/30 between new and old business, but sometimes trends go in other directions despite your best efforts.

I've never understood the sales analogy of 'hunter vs. farmer.' In my experience, it takes a fierce hunter mentality to land and expand inside a customer because it requires the same tenacity as finding new business. Farming, to me, is what Account Managers (AM) do to KEEP the business with existing customers. AEs, by definition, are always hunters, regardless of what stage the project is in. In a perfect Silicon Valley, AEs would extract the full potential from accounts with the help of the SE, and then AMs and CAMs would manage the accounts to ensure they keep renewing licenses or buying products regularly. If only we could all agree on that.

Your biggest challenge will be escalating business quicker than the customer probably feels comfortable doing. There's no two ways about it—they're going to have to spend some money now, and it's your job to make it sting less. This requires writing reports summarizing the data and analytics you've gathered and spinning them into a new sales pitch with an aggressive timeline. Why the rush? Because your job depends on it, plain and simple. In financial terms, the forecasting may be promising, but you're upside down in terms of resources spent vs. revenue. The pilot program revenue (if any) barely covered your travel expenses. Your boss is wondering what's taking so long because they've stuck their neck out to leadership on your behalf, and the CFO and COO have probably built their next two quarters' budgets around the value of your deal. This brings me to one of my favorite topics to rant about: **Pipeline Forecasting.**

I've spent enough time in sales leadership to know how frustrating it is to have a sales team 'blue-sky' their deals by inaccurately reporting the stages of the deal. I've been there, so I know why they do it—to save their asses quarter after quarter. Since most high-tech products are expensive and the sales cycles are longer than most, all you can do is speculate on how you 'feel' the project is going. It's not uncommon to have over 50% of your total business close in the fourth quarter. Some industries still have use-it-or-lose-it budgets that require departments to spend their remaining budget by year-end. It's not terribly efficient, but it's still in practice, especially in SLED accounts.

CRMs like Salesforce have mechanisms in place to prevent 'blue-skying', but the truth is, most sales teams don't use the tool correctly. I've seen companies create rules regarding how salespeople are to report deals in their pipeline, along with every other workaround imaginable to try and get a handle on revenue forecasting. Ask any C-Suite Executive, and they'll tell you that forecasting is the bane of their existence. Nothing will infuriate leadership more than inaccurate revenue forecasting. You may have won your customer's confidence, but if your leadership thinks your numbers aren't trustworthy, you probably won't be in your position for long.

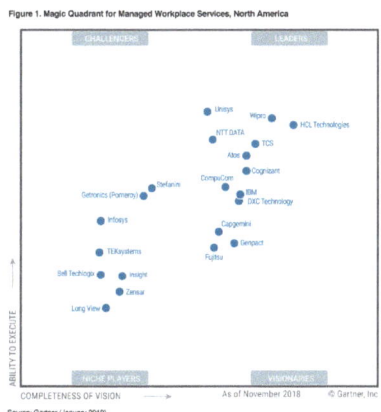

The solution is a very detailed project management report, updated and shared with your leadership regularly. Be as accurate as possible, and put the onus on your direct boss to corroborate the value and likelihood of your pending revenue. This isn't a popular opinion with sales managers because it makes their already difficult job even harder, but it's a more transparent way to operate, and that's what matters at the end of the day.

You may feel like the heat is off now that you've scored a small victory with your customer, but that's an illusion fueled by a false sense of confidence from your leadership. While your boss is telling their boss, *"See, Jane Doe is crushing it on the Rivian Truck account. She already has a pilot program in place, and Rivian's leadership says they 'like the technology"* what they're really saying is, *"See, I told you she was a good hire. Aren't I great for hiring her?"* It's utter nonsense and not a commendable SOP. A better way to report a salesperson's traction with an account is to say, *"Let's look at their project management reports or dashboard, see where they are, and make a forecast based on what we learn."* If leadership needs clarity or more granular results, they can ask for them.

A less efficient alternative is to demand quarterly sales reports (QSRs), which should be quick summaries of project management reports. However, they often turn into tremendous snow jobs by talented salespeople who know how to BS with the best of them. The only useful data I've ever gleaned from QSRs is whether the salesperson was full of it or not. I hate to say it, but AEs like that give hardworking, principled AEs a bad name. I've met my fair share of them, and their careers usually stall after a few years, leaving them in tier 2 companies, making under $150K and complaining about their management.

Back to the hunter analogy: it's time to extract every bit of value from your customer. The best way to *flip* other departments or decision-makers outside the original

SoW is to use your relationships inside the company. In the next chapter, I'll talk about evangelism—creating devotees is as simple as making them drink the Kool-Aid. It's a lot to ask of someone, especially if it doesn't directly affect their department, but you need to illustrate that the tech they love will be far more successful if the entire company gets on board. Point out that competitors will soon learn of their new tech and try to catch up. If only 25% of the customer's company is using your tech, and their competitor implements a 100% solution with a different but similar technology, achieving the same end goal—say, same-day delivery of goods—then all the effort to gain market share will have been for nothing.

Go back to the data and analytics, and drive the message home on the effectiveness of your solution. Make it public if you can, and force their hand. (Always get permission to release white papers if you want to avoid a lawsuit.) If you're still getting pushback on a full enterprise rollout, make sure the resistance is coming directly from the C-level head of the department. I can't stress enough how important it is to keep the ultimate decision-maker in the loop throughout the entire sales process. If you don't, you risk losing the deal at the 11th hour with no apologies from the Chief Officer overseeing your project.

Here's a quick breakdown of who you need to be working closely with, depending on the type of tech: Directly under the C-Suite Executives are typically Vice Presidents and Directors. Whoever you start your first conversation with, make sure you ask who will ultimately make the final decision about onboarding your tech. It's not uncommon to keep the fact-finding and initial exploration of new tech considerations at a lower level. Don't discount the importance of this process or the points of contact charged with evaluating your product. It's highly unlikely you'll meet with the ultimate decision-maker during the first rounds of talks. However, as I stressed before, you need a tactic to loop the C-level executive into the process as quickly as possible.

Here is my strategy for doing so; regardless of how long your product has been available in the market, a decision needs to be made as to what is the value threshold of a deal your company is willing to entertain. This decision is closely tied to what size companies or what verticals you initially choose to target. I am of the mind that emerging technology must adopt a top-down strategy targeting potential customers who can afford a technology solution north of $50k. With that said, I know of software companies who will not entertain a conversation for anything less than $100K and an IT consulting company who lead with a product that has a minimum price tag of $150K. The price of the product must reflect the uniqueness and its disruptive potential. By the very definition of disruptive technology, it is not likely the product is inexpensive unless of course that is what makes it disruptive for instance a free ERP that would be sure to give Oracle and SAP heartburn. To be clear the cost of a subscription or piece of hardware does not have to be expensive, but the overall value of the opportunity must have great potential. The Beam (Figure 3) is only a $1000 for the base model but if Facebook is considering leveraging the technology you could value the total deal at well over $5 Million as their appetite would be enormous. If each of your software seats go for $25 per month it would only take a company committing to 340 users to equal a $100K deal. Times that by 5 years and you're at a half of a million dollars. If your customer is SpaceX and you know there are 800 employees who at a minimum will have to have access to your platform for the next 5 years, the deal value just ballooned to $12M. This is before you sweeten the deal by discounting the monthly subscription to $17.50 per month if they sign a 5-year contract which includes free upgrades, patches and on-boarding bringing the price tag down to $8.4M. What you know that your customer does not is, once your company has a 1000 subscribers, you are going to release an *add-on* module that enhances your products features and that is going to come with a $15 per month price tag. Cha- Ching! You just added a few million to the deal.

The logic behind this strategy is as follows: first, landing a Fortune 1000 logo will significantly enhance your product's credibility and quickly signal to other industry leaders that they should also consider your technology if they want to stay competitive. Second, expenditures over $100K almost always require approval from the head of a department, typically at the C-level. Third, focusing on Tier 1 customers is simply smart business, especially when you consider how much it costs to go to market. It's a waste of resources for an AE with a $220K base salary to spend six months on a project with only a $40K spending ceiling. Nurturing a $40K deal costs the same as nurturing a $10.6M deal, so don't waste your time on small opportunities. MSPs (Managed Service Providers) or VARs (Value-Added Resellers) can handle smaller deals for you all day long if you support them (more on that in the next chapter). Resale partners may even land larger accounts for you, so working within your channels is crucial. Partnering with Solution Providers (SPs) like Gartner can also be highly beneficial. If your product is fortunate enough to be ranked a Leader in Gartner's Magic Quadrant, it means you've achieved the highest composite score for both completeness of vision and ability to execute.

"A vendor in the Leaders quadrant has the market share, credibility, and marketing & sales capabilities needed to drive the acceptance of new technologies. These vendors demonstrate a clear understanding of market needs, are innovators and thought leaders, and have well-articulated plans that customers and prospects can use when designing

their infrastructures and strategies. Additionally, they have a presence in the five major geographical regions, consistent financial performance, and broad platform support." [15]

The fourth and final reason is straightforward: you will grow revenue faster, satisfy investors, and secure your spot at the President's Club retreat on Catalina Island this spring.

I'd rather have a deal be a nonstarter than see a successful pilot program fail. A failed pilot program reflects not only my inability to convert the customer but possibly also the strength of the product's performance in the field. It could also be due to a botched trial account onboarding process that lacked sufficient training and support, making an amazing piece of technology frustrating to use and, therefore, easily discarded.

Don't waste your time on small opportunities when there are bigger fish to fry. Time kills deals, so when it comes to expanding your product's footprint within a company, close the deal and move on to the next opportunity.

---

[15] Magic Quadrants and Market Scopes: How Gartner Evaluates Vendors Within a Market- gartner.com. February 2008.

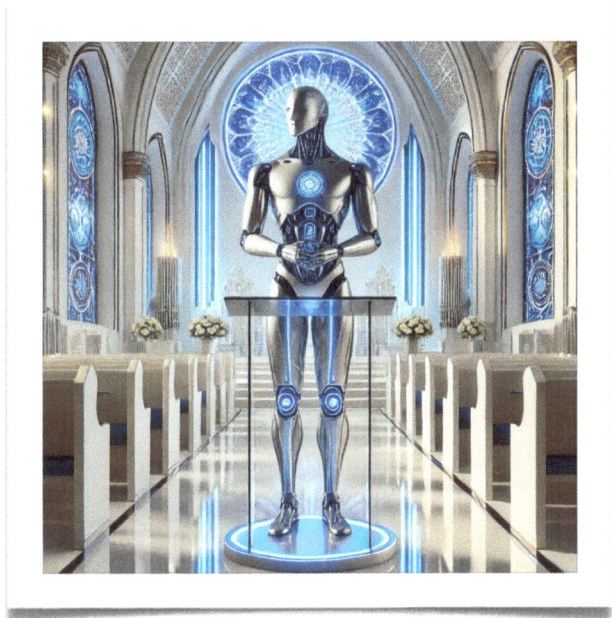

CHAPTER 9 - IF YOUR GOSPEL ISN'T TOUCHING OTHERS, IT HASN'T TOUCHED YOU

I first heard the term *Technology Evangelist* used in conjunction with Guy Kawasaki, Apple's first marketing guru. It resonated with me because it summarized how I felt about certain technology at the time, so I decided right then and there to include it in my professional summary. Needless to say, it confused some people, as the term is typically reserved for marketing professionals. But I didn't care because it defined me as someone passionate about their career and technology as a whole.

Without giving away my age, I can tell you I received my first personal computer, an Apple II, when I was 11 years old. I started coding in BASIC almost immediately, as an operating system hadn't been invented yet. When I was 13, I went off to boarding school armed with a brand-new Macintosh SE. I was by far the nerdiest kid in the eighth-grade dorm and endured more than my fair share of bullying. It wasn't until my classmates discovered they could play games on my computer that I became popular with the technologically challenged. Like Steve Jobs and Richard Branson, I'm dyslexic, so computers as word processors were essential for me to easily edit my papers. To this day, I'd be severely handicapped without that little red line under my misspelled words and Siri's grammar AI. Suffice it to say, I became a born-again Apple Evangelist early on and have been a faithful devotee ever since.

That's not to say I haven't embraced Microsoft, Linux, and the many ISVs who develop for them. Surprisingly, I've never worked for Apple Corporation. Naturally, I heed my own advice and level up my Jedi skills with whatever technology company I work for, so I can tell you quite a bit about search and advertising analytics, SEM & SEO optimization, big data storage, autonomous air cushion vehicles, automated identification capture, enterprise mobile device management, synthetic DNA for nano-identification, radio frequency identification, warehouse and supply chain management software, OKR SaaS, and disruptive e-commerce platforms. I can preach the good word on all these subjects, having sold all of this technology at one point in my 23-year career. Was it all disruptive technology by definition? No, not all of it, but that didn't stop me from becoming a missionary for every technology I've sold.

Evangelism is more than just knowing your technology in a meeting. It's a disciplined practice to establish yourself in the industry as a thought leader. It requires writing white papers, posting regularly on social media and blogs. Your name should become synonymous with your product. It takes time to achieve this, so my advice is to focus on an industry or particular group of technology—like Network Security, DevOps platforms, or Mobile Productivity in Healthcare. This is also great advice for your career. If you stay in the same wheelhouse, you don't have to explain why you shifted from Nano Tech to Mobile Device Security in one move.

In the previous chapter, I suggested turning your customer into a devotee. I'll spare you all the clichés about the power of word-of-mouth, but suffice it to say there's nothing more viral. Having someone on the inside working on your behalf to expand your footprint within an existing customer is the best-case scenario. This is achieved by getting your end user very excited about the product through regular *sermons* on how their work life is forever improved because of your tech. It's a bit like brainwashing, but it doesn't require sleep deprivation and hypnosis.

If you're attending a trade show where you know you have a devotee, make sure they visit you at your booth. Get their permission to give out their email to prospective customers looking for a recommendation. Canonize them in any white paper or public post you make, and stay in close contact long after the deal is done. It goes without saying that you want to build on your successes and leverage your faithful supporters when you start fresh with a new company.

I'm not encouraging you to raise your *recruit me please* flag when you're successful unless you have a very good reason—and no, desiring more money is not a good reason. If you want more money as an AE or SE, work harder inside and outside your company. No one is deliberately stopping you from achieving success, so fix what you can and keep crushing your numbers. If you're successful in your field, recruiters will try to poach you for another company. It usually involves more money, but be careful with your decision—it could backfire if your resume looks like you jump ship for every trending stock offer.

Ever considered starting a church? What would it take to get started saving souls? First, you need a subject of devotion—someone or something to exalt. This is your product or product category, like Autonomous Security Robots. Let's use **Cobalt AI'** product, the Cobalt Security Robot, as an example. It costs $6,000 a month, which is roughly half of what it costs to have three human security personnel patrolling 24 hours a day. So, it does the work of three humans for half the price. Considering that the security industry reports that only half of the available security guard positions are filled, it's a gold-plated solution that, if it gains traction, will greatly disrupt the security industry. [16]

Did I mention it comes standard with 60 sensors, including smoke, O2, infrared, LIDAR, ultrasonic sensors, facial recognition, and AI? It can even be fitted with a Geiger counter if needed. No, it can't mop up spilled coffee, but it can recognize the hazard and call maintenance to clean it up. Don't forget, it's autonomous, so you only need to interface with it when it detects a security situation. Imagine it rolling down the hall at 3 a.m., looking for open doors and sniffing the air, when its infrared sensor notices someone left the toaster oven on in the pantry. Ring, ring—security is notified via phone call, text, or email, and the problem is solved. A few minutes later, it encounters a person it doesn't recognize from

---

[16] Business Wire, Physical Security Market-Increasing Domestic and International Security Threats to Drive Growth | Technavio, February 28th, 2018

its facial recognition database. A Cobalt engineer's face appears on the tablet and asks the person to badge in with their credentials on the robot's RFID reader. They're cleared, and on their way they go—good little robot.

Does this sound like a sales pitch or a passionate evangelist speaking about robotics? It's a little of both. The point is to keep telling the story to anyone who will listen, whether they're a prospect, customer, colleague, or just the person sitting next to you on your flight home to SFO. Who knows? That person might be married to the CISO at Microsoft, and he might call his wife the moment he lands to tell her about it.

As I write this, I'm reminded of the flip-side of creating brands, both personal and ideological. Kurt Andersen addresses this in his New York Times Bestseller *Fantasylan*d: How America Went Haywire: A 500-Year History. In his book, Andersen illustrates how waves of Americans have been drawn to the New World because it was a place they could create their own realities, with little objectively regulated truth standing in their way. Creating a brand of yourself is necessary to some degree in order to get hired for a job. Let's agree it's a tightrope we have to walk between being an *expert* and becoming a *brand*. So, let's try to keep our role in the grand scheme of things in perspective. With that said, you are a product yourself. At some point in the sales process, you're going to have to ask your customer to buy into the fact that your involvement in the project is essential for its success. The old adage is true: people buy from people they like.

What does your resume or LinkedIn page say about you in the first 20 words? Hopefully, you have a career summary at the top of your page. Mine says I wrote this book and that I'm a thought leader on a number of technologies. My name is synonymous with disruptive technology today. It doesn't take much to throw your hat in the ring before journalists start writing to you for comments, and local business guilds invite you to speak. If you're gifted at promoting yourself via information channels, get to it. It's hard to keep up with the amount of content necessary to set yourself apart, so I'm considering hiring someone to be my social engineering agent. Oh, it's a thing. If that gets traction, I could also hire a business agent to get me more public speaking gigs, and maybe the coveted TED Talk gig in Vancouver. Local chapters are fun, but Vancouver and New York TED Talks are the *big shows*.

I bring personal ambition up because this is the mindset you need to adopt to be successful in tech. Not only are you a skilled AE/SE/AM with an exciting career, but you are also a commodity—not unlike the tech you're currently selling. This means your public profile needs to be consistent with how you wish to position yourself to your customers. You just walked into a room and claimed, in so many words, that you are a Jedi Knight with an impressive understanding of your client's business. Will a quick Google search contradict this? Let me put it this way, Skywalker: does your Facebook, Twitter, GitHub, and LinkedIn support your claims of being an ordained Jedi? Are there white papers and industry reports with your name on them? If not, it's time to turn on the content generation machine and get to work.

Honestly, if you haven't updated your CV to include a comprehensive skill set that covers everything from CRMs and task management suites to cutting-edge AI tools like ChatGPT and machine learning-powered business analytics platforms, you're setting yourself up for disappointment. In today's tech-driven world, where innovation moves at breakneck speed, simply knowing how to navigate basic tools like Excel and Office 95 is the equivalent of showing up to a Formula 1 race with a horse and buggy. Hiring managers at bleeding-edge tech companies aren't just looking for someone who can check off a list of outdated skills—they're searching for candidates who display an eagerness to adopt and master the latest technologies.

Resourcefulness and adaptability are no longer optional; they're non-negotiable. You wouldn't expect a carpenter to show up at a job site without a hammer or a saw, so why would you expect to impress a tech company without demonstrating that you're equipped with the modern tools necessary to thrive? And let's be honest—if you're not keeping pace with innovations, your competition will. At the risk of over-explaining this, don't bring an Amish horse-drawn plow to the Googleplex. Instead, come prepared with a metaphorical toolbox full of AI-driven insights, automation capabilities, and next-gen problem-solving skills. It's not about knowing everything right now, but showing that you have the drive and the tools to learn fast and keep up with the industry's rapid evolution.

Landing and expanding within a client's organization is about much more than simply closing a deal. It's about embedding your solution so deeply into their operations that their end users become loyal advocates—people who not only use your product but become vocal champions, advocating for its value across the company. Turning end users into devotees, and ideally evangelists, is key to scaling your footprint within a business. This creates organic momentum, where the buzz about your solution spreads naturally, making it far easier to expand to other departments or divisions.

One particular experience that underscored this for me was a deal with a global prepared foods company. The project involved rolling out our solution across 26 factory locations, spread over four separate divisions throughout the U.S. On paper, this was a major opportunity for growth. However, the challenge I faced was that my primary point of contact—the Regional VP—had authority only over a specific geographic area and lacked the decision-making power to influence corporate-level executives, particularly the CTO, who was the ultimate gatekeeper for tech adoption across the entire enterprise.

This created a significant hurdle. While the Regional VP was enthusiastic and had a clear understanding of how our solution could drive operational efficiencies in his region, his influence didn't extend to the broader company leadership, who were ultimately responsible for long-term strategic decisions. It quickly became clear that a successful expansion would hinge on more than just a strong implementation at the factory level; it required us to win over stakeholders much higher up the corporate ladder.

Our approach had to be multifaceted. First, we focused on creating quick wins with the regional factories—demonstrating immediate, tangible value that the VP could take to his peers and superiors. We also trained the operational teams at each factory extensively, ensuring they became proficient users who saw our solution as indispensable. These end users became key advocates, feeding success stories back to leadership, which helped build internal credibility.

At the same time, we began crafting a strategy to appeal directly to corporate leadership. This involved aligning our solution with the broader objectives of the company—particularly those of the CTO. We framed our product not just as a regional efficiency booster but as a strategic tool that could be scaled across all divisions to support the company's larger technological goals. By showcasing how our platform integrated with their existing systems and helped future-proof their supply chain, we positioned it as essential to their overall digital transformation efforts.

The combination of bottom-up and top-down influence created a flywheel effect. As the operational teams and the Regional VP continued to champion our solution, the corporate leadership began to take notice. Bazinga! Eventually, the CTO approved a pilot at several additional factories, and once the initial results demonstrated the kind of impact we'd been promising, we were able to secure a broader rollout across the company.

In the end, it wasn't just about winning over a single decision-maker. The true challenge was converting end users into advocates, leveraging those advocates to build credibility internally, and ultimately positioning our solution as a strategic imperative for the broader organization. This experience reinforced for me that true client expansion depends on building influence at all levels of an organization—from the factory floor to the C-suite.

.

CHAPTER 10 - FALL SEVEN TIMES AND STAND UP EIGHT

As Japanese proverbs go, this is one of my favorites. It's a simple illustration of what success looks like. Building on your success is the subject of the final chapter of this book. **The Paradox of Disruptive Technology** can be countered by a growing reputation of successful implementation. Industry is inherently competitive, so success stories travel far and wide at, unsurprisingly, the same speed as failures. Both insights are valuable when sizing up the market. However, the outcome of your deal, if it's with an important enough customer, will get others to take notice.

If it's a success, and you're growing your business with the customer year over year, capitalize on this and use it as a springboard into new business with similar needs. At some point, your tech will become a *must-have* tool that everyone demands. As dynamic as SAP is as an ERP platform, there was a time when no one really knew what it could do or improve, but everyone had to have it. It must have been amazing being an AE for them back then. If GM went with SAP, then Ford and Chrysler bought it sight unseen. Technology was evolving quickly in the 90s, so to stay relevant, companies were gobbling up SaaS with barely a second thought—sometimes not even using the platform but paying for it just in case they had to. It was madness, and I see this happening again with data security and DevOps. Companies are moving quickly to adapt and survive, so if your tech is the new black, prioritize your customer list and decide which new opportunity will bring you the most revenue and exposure. Work from the top down and try to grab as much revenue as you can while you can. The faster your company exceeds its revenue goals, the quicker your fistful of stock options will be worth their weight in bitcoin (if blockchain had any mass, which it doesn't).

We should probably address failures while we're on the topic. I don't know who first said *fail quickly*, but it's one of those business-isms I love to say to young Biz Devs when we're out for drinks after a trade show. When your deal goes south, try to figure out why, so you can learn from the experience. I believe there are only two reasons deals ever die in the late stages of the sales cycle: the AE dropped the ball, or the customer had an unforeseen change in course. These are pretty loaded statements, so I'll break it down for you.

If the salesperson dropped the ball, that could also mean they lost the sale to the competition (if there is any). It could also mean you weren't likable or trustworthy. The worst scenario is if your product didn't deliver. There are many ways to screw up a deal, so figure out what you could have controlled and forget the rest. If you have a 50% close rate in any industry, you're doing pretty well, but those deals better have been nonstarters or failed early. If your product didn't get through testing, don't blame the product. These things happen, but in my experience, it should be the exception to the rule and very rare. I say that because your testing doesn't always have to be a success the first time. However, if you're dug in and trusted by the end users, you can still push the deal forward with new developments and future testing. So, failure can be turned into success if you manage the project effectively.

This brings us full circle to the simplicity that has always guided my approach to selling and delivering emerging disruptive technology. It's a straightforward yet powerful roadmap: make your first appearance on the scene with confidence and make an impression that sticks. Manage each project with unwavering consistency and transparency, using a best-in-class platform that you and your clients can rely on. Then, as success builds, capitalize on that momentum by landing and expanding within your customer base, maximizing revenue at every opportunity. Finally, leverage that success to lay the foundation for future growth. While this plan may seem simple, it's mastering the subtleties of navigating the Paradox of Disruptive Technology that will ultimately define your success. It's in those finer details—the nuances of timing, market readiness, customer adoption, and scalability—that the real work and true rewards lie.

As you step into this ever-expanding world of technology, I sincerely hope that your journey is as fulfilling as mine has been over the past few decades. Don't hesitate to step into a room, confident in your knowledge and expertise, and be ready to position yourself as the go-to authority in your specific technology domain. Confidence is key, but so is clarity. It's not enough to simply know that your product is disruptive—you need to clearly and compellingly demonstrate how it will deliver tangible, measurable benefits to your customers. Help them understand how this new technology will solve their pain points, streamline their operations, or unlock new opportunities. Your ability to translate technical jargon into real-world value is where the magic happens.

Always remember, data is your ally. Support every claim you make with solid evidence—reliable data, insightful analytics, and results from successful testing. Pilot projects, case studies, and early wins are invaluable for showing your customers that you're not just selling hype; you're delivering proven outcomes. And never underestimate the power of community backing. Surround yourself with strong partners, investors, and advocates who believe in your vision and the potential of your technology. That external validation can provide the momentum you need to push through challenges and scale your solution.

But through it all, don't forget to enjoy the process. There's a unique joy in watching a project come to life—from the first concept to the final deployment. There will be hurdles, of course, but the sense of accomplishment when your technology moves from an idea to something that truly impacts businesses or industries is unlike anything else. Take pride in every milestone, and remember to celebrate the small victories along the way. You have a lot to be proud of already, and that feeling will only grow as you continue to innovate and succeed.

When you look back on your career, whether it's months or decades from now, you'll have the satisfaction of knowing you played a critical role in building something meaningful. You'll be able to reflect on your journey with a deep sense of pride, knowing that you didn't just witness technological change—you actively contributed to it. You helped shape the future. And when you can confidently say, "*I helped build that*," you'll know that your work has made a lasting impact on the world around you. That's a legacy worth aspiring to and celebrating.

~Fin~

2024 APPENDIX

Since the initial release of this book back in 2019, the world of technology has undergone a seismic shift. What we're witnessing today is nothing short of a another tech boom—innovations that were once theoretical have become mainstream, and new breakthroughs are emerging at a dizzying pace. This isn't just about the usual improvements in processing power or incremental software updates. We're talking about the arrival of truly disruptive technologies that are rewriting the rules across industries.

Take Edge Computing, for example—what once seemed like a niche concept has now become essential, bringing processing power closer to the source of data to minimize latency and maximize efficiency. It's transforming industries from manufacturing to healthcare, offering solutions that were simply impossible a few years ago. Then there's Distributed Ledger Technology (DLT)-as-a-Service, a game-changer for businesses looking to integrate blockchain-based solutions without building them from scratch. The rise of DLT is simplifying everything from supply chain management to finance, creating new business models along the way.

And of course, we can't overlook Web3, the decentralized vision for the internet's future. Web3 is empowering users to take back control from the tech giants, offering new possibilities for data ownership, digital identity, and peer-to-peer transactions. It's not just a buzzword—it's a movement that's set to redefine how we interact online.

But perhaps the biggest players in this evolving landscape are Artificial Intelligence and Machine Learning. These once-sci-fi technologies are now so integrated into our lives that we often don't even realize it. From personal assistants that learn our preferences to complex algorithms optimizing entire supply chains, AI and ML are everywhere. Love them or loathe them, their impact is undeniable—they're the supervillains and superheroes of modern tech, depending on your perspective.

And while all of these advancements are reshaping the present, there's one technology on the horizon that promises to disrupt everything we know: quantum computing. Quantum computing is poised to be the ultimate game-changer, far beyond anything we've seen with AI, DLT, or even Edge Computing. Harnessing the strange and powerful laws of quantum mechanics, these computers are set to solve problems that would take traditional systems centuries to crack. From unbreakable encryption to revolutionizing drug discovery and optimizing complex systems like never before, quantum computing has the potential to redefine entire industries overnight. It's not just an incremental step forward—it's a leap into an entirely new era of computing. While we're still in the early days of its development, the race is on, and those who can grasp its implications and prepare for its inevitable impact will have a serious competitive edge in the tech landscape of tomorrow.

Recognizing the immense impact of these developments, I've expanded this book to include new chapters that dive deep into each of these disruptive forces. I explore not just what they are, but why they matter, and how you can strategically position yourself to capitalize on them. Whether you're an entrepreneur, a tech professional, or simply someone who wants to stay ahead of the digital curve, these new sections are packed with insights and practical advice to help you navigate this brave new world.

So, if you're looking to be the trailblazer in your industry, the first on your block to sell these groundbreaking tech stacks, this expanded edition is your key to unlocking the future. From theoretical frameworks to real-world applications, it's all here, laid out in a way that's accessible and actionable. Stay ahead of the curve—because in the fast-moving world of tech, those who stand still risk being left behind.

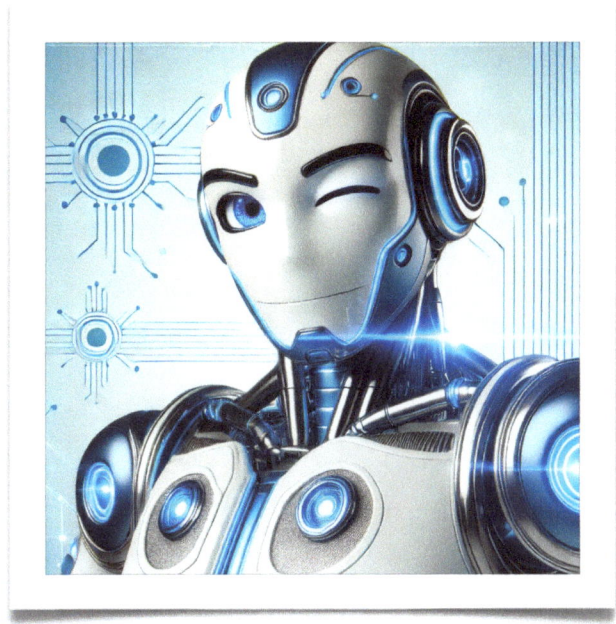

1: ALWAYS SAY PLEASE AND THANK YOU TO YOUR AI, OR ELSE...

If you think AI and machine learning (ML) are just fancy buzzwords, it's time to buckle up—because this stuff is about to blow your F'in mind. Let me put it this way: AI and ML are the most disruptive technologies to hit humanity since electricity lit up our nights and revolutionized, well, everything. If that sounds like a bold statement, it's because it is. But stick with me, and I'll explain why AI/ML is not only the present but the future.

Imagine the world before electricity—dark, slow, and full of people squinting at things in candlelight. Now imagine the world without AI and machine learning. That's right, it's hard to picture, because AI is already everywhere. It's in the apps we use, our smart phones, the websites we browse, even the cars we drive (or soon won't have to drive). And like electricity, it's going to keep expanding into every corner of our lives, making things faster, smarter, and way more efficient.

Let's get one thing straight: AI isn't just robots or some dystopian sci-fi future where machines take over (although, who knows, maybe someday). AI is already a massive part of our everyday lives. It's making decisions, solving problems, and doing tasks we used to rely on humans to do. And it's doing them better, faster, and at a scale we could only dream of.

Machine learning—AI's slightly nerdier sidekick—is the technology that allows machines to learn from data. Think of it like this: electricity powered machines and appliances, but AI/ML powers smart machines that *think* and *learn*. It's what lets your Netflix account suggest the perfect show based on your weird binge habits, or your phone finish your sentences (sometimes a little too accurately, right? i.e duck a duck). And here's where it gets wild: AI/ML is doing all of this without any hand-holding. The more data you feed into these systems, the better they get. It's like having a personal assistant who reads your mind, anticipates your needs, and becomes more competent the more they work for you—minus the awkward "*we need to talk about your attitude*" moments.

Let's flashback to when electricity first came along. It wasn't just a cool way to light up rooms. It changed everything—how we lived, how we worked, how we moved, how we connected with one another. Factories sprang up overnight, homes became modernized, and the way humans interacted with the world fundamentally shifted. AI and ML are doing the same thing, but instead of powering machines, they're powering decisions.

The best part? We're only scratching the surface of what AI/ML can do. We're talking about automating repetitive tasks (adios, data entry), predicting trends, diagnosing diseases with pinpoint accuracy, and even driving cars. It's changing how industries function, making them faster, smarter, and leaner. Just like electricity revolutionized everything from factories to homes, AI is revolutionizing everything from healthcare to finance.

Consider healthcare for a moment. AI is already diagnosing conditions faster than doctors, helping researchers develop new drugs, and optimizing hospital operations. It's even helping predict outbreaks and pandemics (yeah, we could have used a little more of that foresight in 2020). The healthcare industry is getting a serious AI upgrade, and we're all going to benefit from it.

Now, think about finance. Gone are the days of human stockbrokers furiously scribbling on notepads. Today, AI algorithms are trading stocks, analyzing markets, and making financial decisions faster than any human ever could. In fact, AI is so good at spotting trends and patterns that it's become Wall Street's secret weapon. And yes, it can even help you avoid overdrawing your account, which would have been nice during my college years.

AI isn't some futuristic tech that's five years away. It's already here, and it's already making waves. From voice assistants like Siri and Alexa to recommendation algorithms on YouTube, AI is everywhere, subtly improving your life. Well, mostly. Let's not forget that time Alexa laughed out of nowhere in the middle of the night—that was creepy.

Machine learning, the engine behind all of this, is what makes AI so powerful. Every time you use an AI-powered tool, you're teaching it. That's right—every playlist Spotify creates for you, every product Amazon recommends, and every meme Instagram serves up is based on your behavior. The more you interact, the better the AI gets at guessing what you want. And that's kind of amazing.

But AI isn't just about making your online shopping experience more pleasant. It's solving real-world problems. For instance, AI is being used in agriculture to monitor crops, optimize water usage, and even detect disease early—before it wipes out an entire field. Farmers are using AI-powered drones to monitor soil conditions and predict weather patterns, making food production more efficient. So yeah, you can thank AI for those perfectly ripe avocados.

But let's not kid ourselves—AI isn't all rainbows and unicorns. Just like electricity had its share of risks (hello, electrocution), AI has its downsides too. Privacy is a big concern. With AI getting smarter, it's gathering more data about us than ever before. From our shopping habits to our sleeping patterns, AI is always watching—and that can feel a little invasive.

There's also the question of jobs. With AI automating so many tasks, what happens to the humans who used to do those jobs? The good news is that while AI will definitely replace some jobs, it will also create new ones—ones we can't even imagine yet. Think about it: when electricity became mainstream, it didn't just replace candle makers. It created entirely new industries—electricians, appliance manufacturers, and eventually, software developers. The same will happen with AI.

But the big question? How do we make sure AI doesn't turn into the villain of the story? The answer lies in responsible development. It's up to us to set the rules, establish ethical guidelines, and ensure AI works *with* us, not against us. After all, we don't want a Skynet situation on our hands, do we?

If you think AI/ML has already peaked, think again. We're in the early stages of this revolution, and the potential is limitless. Industries are still figuring out how to fully harness the power of AI, and as they do, we're going to see a seismic shift in how the world operates. Everything from manufacturing and healthcare to entertainment and education will be transformed.

Imagine schools using AI to create personalized learning experiences for every student, hospitals using AI to predict outbreaks before they happen, or cities using AI to optimize traffic flow and reduce pollution. The possibilities are endless, and we're just getting started. In the future, AI might not just be the new electricity—it might be the new air we breathe. Invisible, but essential to everything we do.

So, what does all this mean for you? It means AI and ML aren't just the most disruptive technologies since electricity—they're the most exciting ones too. If you're not already diving into AI, now's the time to jump in. Whether it's improving business processes, solving real-world problems, or just making your life a little easier (thanks, Siri), AI is here to stay—and it's only going to get better.

I'm often asked by young graduates considering a career in tech how to get started. I usually respond with a question: *"What are you good at? Do you like math?"* If so, coding might be for you. *"Do you enjoy talking to people?"* Consider sales or business development. *"Are you creative?"* Then marketing, especially where it intersects with cutting-edge tech, could be your path.

But today, with AI being the biggest advancement since Al Gore *invented* the internet, I encourage driven young professionals to carve out a role in designing and maintaining AI toolsets for startups and enterprises. The ultimate goal? Becoming one of the first Chief AI Officers (CAIO). This kind of role could command a salary with on-target earnings (OTE) exceeding $250K!

You won't necessarily need a degree in robotics, electrical engineering, or even coding. Instead, you'll need to understand what tools like ChatGPT, Google's NotebookLM, or the many other AI-powered platforms—those that generate videos, images, podcasts, or customer support bots—can do to boost efficiency and save companies money through automation. By being the master behind the curtain of AI/ML platforms, you'll create a job that's becoming indispensable—with minimal competition, for now.

So embrace the disruption, stay curious my dudes, and keep an eye on what's coming next. Because trust me, it's going to be a wild ride.

2: LEDGER TECHNOLOGY - THE OLDEST TECH IN THE WORLD

Distributed Ledger Technology (DLT) has got a bad rap these days with all the founders going to jail and no end to the token rug-pulling scams every day. The truth is, DLT is one of the coolest (and most useful) innovations out there today. It's the technology behind blockchain, but here's the kicker—it's *so* much more than that. Think of DLT as blockchain's more flexible, more talented cousin, and now it's available as a service. Yep, welcome to **DLT as a Service** (DLTaaS), where you don't need to be a cryptography genius or spend a fortune building infrastructure from scratch. You just subscribe, like Netflix, and boom—you've got yourself some cutting-edge distributed ledger action.

Ever since man could scratch hash marks on cave walls, we have been perfecting ways to keep track of our stuff. Ledgers are profoundly essential to societies and constitute the oldest technology spare the cultivation of fire.

Why is DLT-as-Service the shiny new thing that everyone's talking about? Well, for starters, it's all about transparency and security. We're living in a world where data breaches are more common than your morning coffee. DLT, however, keeps things locked down tight. It's like an un-hackable diary that everyone can read but no one can tamper with. And it's not just about keeping your secrets safe—it's about making sure *everyone* plays by the same rules. Plus, DLTaaS is customizable and scalable, which means businesses of all shapes and sizes can jump on board. Whether you're running a supply chain, a healthcare company, or a financial services firm, DLTaaS has something to offer. Imagine being able to track every product in a supply chain with absolute certainty, or having real-time access to medical records without worrying about hackers. Morpheus.Network is one such solution that is quickly becoming a must have platform if you have a complex global supply chain. That's the kind of magic DLTaaS brings to the table. And here's the fun part—real-time auditing. You know all those long, boring audits that require sifting through piles of paperwork? Well, with DLTaaS, you can pull up a transparent audit trail with just a click. Goodbye, endless spreadsheets. Hello, instant compliance!

Now let's get into the fun stuff—how DLTaaS is making the world a better place (okay, maybe that's a stretch, but it's definitely making it more efficient). Companies like Walmart and IBM are already using it to keep track of their supply chains. Imagine being able to follow a product's journey from factory to shelf without the fear of fraud or missing shipments. It's like having a superpower for logistics.

In finance, DLTaaS is changing the game with cross-border payments and digital assets. No more waiting days for a transaction to clear or paying ridiculous fees to move money across borders. It's all instant and cheaper—what's not to love? And in healthcare, DLTaaS is a godsend for patient records. No more misfiled paperwork or outdated records. With everything securely stored on a decentralized ledger, healthcare providers can access updated information instantly. It's like a super-efficient doctor's office in the cloud. Even governments are getting in on the action. DLTaaS is being used for everything from voting systems to land registries, making sure everything is transparent and corruption-free. Imagine a voting system you can actually trust—pretty disruptive, right?

Okay, so as much as I love DLTaaS, it's not without its challenges. The big one? Regulations. Governments haven't quite figured out how to regulate cryptocurrencies, let alone distributed ledgers. And if you're integrating DLT into an old, creaky system, well, good luck with that. It can be a bit like trying to fit a square peg into a round hole—it's going to take some effort.

Then there's the issue of scalability. Some DLT systems, especially blockchains, can be a bit slow. You don't want to be the person holding up the line because your ledger is taking forever to update. But fear not—tech geniuses are working on new solutions, like Proof of Stake, to make DLT faster and more efficient. But let's not get bogged down by the challenges—because the opportunities far outweigh them. From tokenized assets to digital currencies (yes, I'm talking about you, Central Bank Digital Currencies), DLTaaS is opening doors we didn't even know existed. Whether you're in gaming, healthcare, or supply chain management, DLTaaS has the potential to shake things up in the best way possible.

So, what's on the horizon for DLTaaS? *Interoperability*, my dude! Right now, there are a bunch of different DLT systems that don't really talk to each other. But the future will bring systems that play nice together—think blockchain hanging out with DAG and Hashgraph at a tech party. And then there's AI integration. Imagine DLT paired with a decentralized artificial intelligence agent like SingularityNET—that not only record data but also predict trends and automate decisions. The possibilities are endless.

We've also got to talk about going green. The traditional blockchain gets a lot of flak for using crazy amounts of energy, but the tech world is shifting towards greener solutions, like Proof of Stake. So, no need to worry—DLTaaS will be as eco-friendly as it is efficient. Oh, and let's not forget quantum computing. This might sound like something out of *The Matrix*, but quantum computers could theoretically crack the encryption that keeps DLT secure. Don't panic though—DLT providers are already working on quantum-resistant cryptography, so we'll stay a step ahead of the machines. (Cue *Terminator* theme.)

Look, if you take one thing away from this chapter, it's this: DLTaaS is the future. Companies that jump on the DLTaaS bandwagon now are going to have a massive edge over the ones that wait around. It's like the early days of cloud computing. Remember those companies that hesitated? Yeah, they had to scramble to catch up while the early adopters reaped all the rewards. It doesn't matter if you're in finance, healthcare, logistics, or even government—DLTaaS is going to change how we do business. And the best part? You don't have to build the system yourself. Just subscribe, like Netflix, and let DLTaaS do its thing. In a few years, we're all going to look back and wonder how we ever did business without it. So, the real question is: Are you going to be one of the pioneers who helped shape the future, or are you going to sit on the sidelines and watch everyone else pass you by? Your move.

3: HACK THIS MF!

Imagine this: you're lounging on your couch, scrolling through your phone, watching your smart fridge tell you you're out of milk. Life is good. But behind the scenes, there's a battle raging—hackers trying to break into databases and steal your personal info. We've all seen the headlines, right? **Major Data Breach Exposes Millions!** It feels like it's happening all the time. But guess what? There's a new hero in town that's ready to change the game and keep those hackers at bay: Edge Computing.

Now, if you've never heard of Edge Computing, don't worry—you're not alone. It sounds like something James Bond might use to save the world, but it's actually a technology that's about to become your new best friend. In short, Edge Computing is about processing data closer to where it's being generated, rather than shipping it all off to the cloud or some central server. It's like bringing the power of a supercomputer right to your doorstep, or, more accurately, to your devices.

So, why should you care about Edge Computing? Well, let's start with the big one: **security**. Think of traditional data storage like a massive vault. All your data—everything from your banking details to your smart thermostat settings—is stored in one place. And while that vault might have some serious locks on it, it's still a single target. Hackers love this. If they can crack that vault, they hit the jackpot.

Enter Edge Computing. With edge tech, your data isn't all stored in one central location anymore. Instead, it's processed and stored locally—right on your device, or at least much closer to you. This means that if a hacker wants to get their grubby hands on your data, they can't just hit one central server and steal it all. They'd have to break into thousands of different m*ini vaults*, each of which has its own defenses. It's kind of like the difference between trying to steal the crown jewels from a heavily fortified castle versus trying to rob every house in the entire city at once. Not so easy, right?

The decentralized nature of Edge Computing means that even if one device gets hacked (like your smart fridge, because who wouldn't want to hack your grocery list?), the damage is limited. The rest of your data, which might be stored on your phone, laptop, or local edge servers, remains safe and sound. Hackers now have to work a lot harder, and honestly, most of them are lazy. They'd rather move on to easier targets.

Now, we're all living in a world where *smart* things are everywhere. Smart TVs, smart speakers, smart lights, smart thermostats—the Internet of Things (IoT) is growing faster than you can say *"Alexa, turn off the lights."* But here's the thing: every one of these connected devices is a potential target for hackers. And the more devices you have, the more potential vulnerabilities you're adding to your life.

This is where Edge Computing really shines. By keeping data processing and storage closer to where it's actually being used (like right on your smart TV or smart speaker), Edge Computing reduces the need to send data back and forth to the cloud all the time. That means less exposure to the big, bad internet where hackers like to hang out, waiting for an opportunity to strike.

Think of it this way: every time your smart light sends a signal to a faraway cloud server to figure out whether you want it dimmed or turned off, that's a little moment where your data is vulnerable. With Edge Computing, that decision is made right on the device or a nearby edge server. It's faster, safer, and your data doesn't have to take a dangerous trip across the wilds of the internet.

So, we've established that Edge Computing makes it harder for hackers to do their dirty work, but how does this make *your* life safer? Let's break it down.

First off, with Edge Computing, the chances of a major data breach are way lower. If your bank, for instance, is using Edge Computing, it's storing and processing sensitive data closer to where the transactions are happening—right at the ATMs, or even on your banking app itself. So, instead of hackers breaching a central server and accessing millions of accounts at once, they'd have to individually target each ATM or app session. Spoiler alert: that's not happening.

Second, Edge Computing can help protect critical infrastructure—stuff like power grids, water systems, and transportation networks. These are the things we civilians rely on every day without thinking twice. But guess what? They're prime targets for hackers. By implementing Edge Computing, these systems can become much more secure, because the data is being processed at local nodes, rather than one giant hub. This makes it way harder for hackers to cause widespread chaos.

And lastly, Edge Computing is set to revolutionize things like autonomous vehicles, drones, and smart cities. Imagine a future where traffic lights communicate with your car, adjusting based on real-time traffic data to avoid jams. Or drones delivering packages while avoiding power lines, thanks to local processing of sensor data. These things can't rely on sending data back and forth to a distant cloud. They need to make decisions instantly, on the spot. And the less data that's floating around in cyberspace, the safer you'll be from someone hijacking your ride or drone.

Hackers don't like Edge Computing, and it's easy to see why. It takes away their central target and forces them to work much, much harder. With traditional systems, they could hack into one database or one cloud server and walk away with a treasure trove of data. But with Edge Computing, they're faced with a fragmented system. They'd have to break into thousands—sometimes millions—of edge devices to get the same amount of data. That's a ton of work, even for the most persistent hacker. Plus, with data being processed and stored closer to the source, there's less data being transmitted over the internet, which means fewer chances for hackers to intercept it. It's like trying to eavesdrop on a conversation that's happening in a soundproof room—you're going to have a bad time.

Oh, and here's another kicker: Edge Computing is faster. Processing data locally means you don't have to wait for it to travel across networks and back again. Whether it's your car making a split-second decision or your security camera alerting you to a visitor, the speed of Edge Computing means safer, more responsive systems. And that makes our everyday lives not only more secure but a whole lot smoother.

We're heading toward a world where Edge Computing will be as common as Wi-Fi. In fact, it's already happening. As 5G networks roll out and the IoT continues to explode, Edge Computing is going to be the backbone of our digital future. And the best part? It's making that future a whole lot safer.

You'll still get all the convenience of smart devices, but with an added layer of protection against the hackers lurking in the shadows. And while the fight against cybercrime will never be over, Edge Computing is a powerful new tool in our arsenal—a tool that's making it harder for the bad guys to win.

So, next time you ask your smart speaker to play your favorite song or turn off your lights, just remember: somewhere in the digital ether, Edge Computing is working quietly behind the scenes, keeping your data safe, your devices secure, and your life just a little bit safer.

4: WEB3 OR BUST! (AND WHY BURNING MAN IS SO 20 MINUTES AGO)

Let's talk about Web3—the internet's glow-up that's going to change the game when it comes to privacy, identity, and how we exist online. Right now, you probably feel like your personal data is scattered all over the internet like confetti after a parade. And who's sweeping up all those digital breadcrumbs? Big tech. Companies like Facebook, Google, and Amazon have been following you around for years, piecing together everything from your favorite pizza toppings to which cat videos you binge at 3 a.m. It's a little creepy, right? But here's the thing: Web3 is about to shake things up in the best way possible. It's like we're finally getting our digital privacy back. And I'm not talking about installing an ad blocker and calling it a day. I'm talking about true *anonymity*—

where you can browse, shop, and interact online without constantly feeling like someone's peeking over your shoulder.

Trust me, if Web3 were a fashion trend, it'd be the **new black**. It's sleek, it's powerful, and everyone's trying to get a piece of it. If you're still riding the wave of Web2 and thinking that decentralized tech is just for blockchain fanatics, well, it's time to update your closet—just like Burning Man is so 20 minutes ago, Web3 is the shiny new thing all the cool tech kids are into. It's the future, and if you want to keep up, you better grab your virtual reality headset and your cryptographic keys, because this is where it's all happening.

Before we dive into how Web3 is going to give us back our anonymity, let's clear up one thing: Web3 is basically the next evolution of the internet. Think of it like this: Web1 was the early days of the internet (think static web pages and AOL dial-up tones), Web2 is where we are now (social media, user-generated content, algorithms watching your every move), and Web3 is the new frontier. It's the Wild West of decentralized systems, blockchain technology, and, most importantly, *you* owning your data. Here's the mantra of Web3: **Own your keys, own your identity.** What does that mean? Well, it's all about putting the power back in your hands. In the world of Web2, you don't really *own* anything. All those social media profiles, email accounts, and shopping habits you've got stored online? They're sitting on somebody else's server. In Web3, you get to control your own identity with cryptographic keys—no middleman required.

These keys act like a digital safe that only you can unlock. Want to sign into a website? No need to hand over your email or create another password you'll forget in five minutes. You'll use your private key to prove who you are without revealing your personal info. It's like having a secret handshake with the internet. *Big tech can't peek behind the curtain anymore*—your identity stays safely tucked away, and you can still do all the things you love online without worrying about your data being hoarded like a dragon's treasure.

Let's talk about the elephant in the room—*identity theft*. In today's online world, identity theft is easier than ordering a pizza. One major data breach and suddenly some random hacker in a basement is buying a hot tub with your credit card. But in the world of Web3, that's going to get a whole lot harder. Why? Because your identity is decentralized, stored in a way that hackers can't just break into one big database and steal everything in one go.

In Web3, your digital identity is spread out over a secure network (thank you, blockchain!), so there's no one place for hackers to attack. It's like trying to steal the crown jewels, but the jewels are scattered in a thousand different vaults, each one with its own security. Good luck with that, hackers! And the best part? Since your data isn't floating around in a centralized database, your personal details won't be attached to the websites you visit or the products you buy. You'll have a unique identifier—kind of like an anonymous username—that lets you do your thing online without broadcasting your actual identity to every corner of the internet.

Now, I can hear you asking: "*But what about personalized content? I like getting targeted ads for products I might actually want to buy!*" Don't worry, Web3's got you covered there, too. Just because you're anonymous doesn't mean you have to give up the convenience of personalization. In fact, Web3's going to make it even *better*—without selling your soul (or your data) to big tech.

Here's how it works: In Web3, companies can still target you with personalized ads, but they don't need to know *who* you are. Instead, they'll be able to target your preferences—your "buying persona," if you will—without ever attaching it to your real identity. Think of it like this: they'll know you're into organic coffee and hiking gear, but they won't know your name, email, or which dog park you go to on Sundays.

It's all thanks to something called **zero-knowledge proofs**. Stay with me here—it's not as complicated as it sounds. Zero-knowledge proofs let you prove that you meet certain criteria (like, I'm into yoga mats) without revealing any additional information. So, companies can still serve you up those perfectly curated ads, but they don't get to see any of your personal details. It's like waving your hand and saying, *"Yeah, I'm your target audience,"* without handing over your driver's license. Now, I know what you're thinking: What about big tech? How are Facebook and Google going to survive if they can't spy on everything we do? Well, don't worry about them—they're resourceful. They'll still be able to target you with relevant content, but without spying on you like they used to. Instead of hoarding your data, they'll have to adapt to this new world where they play by our rules.

You'll still get ads for that portable blender you've been eyeing, or recommendations for shows that match your late-night binge habits, but they won't know it's *you* specifically. They'll just know they're talking to an anonymous user who likes action movies and orders pizza on Fridays. It's personalized without being invasive, which is a win for everyone.

Think about it like being at a masked ball—everyone's dressed up and interacting, but no one knows who anyone else really is. You still get to enjoy the party, but your identity stays hidden behind your mask. In Web3, you get to own your data, protect your identity, and still enjoy all the perks of a personalized internet without worrying about your privacy being compromised.

So, here's the bottom line: Web3 is about to hand us back something we lost a long time ago—control over our own identity. Instead of big tech watching our every move and selling our data to the highest bidder, Web3 puts us in the driver's seat. We'll be able to browse, shop, and interact online without leaving a trail of personal information for hackers (or big corporations) to pick up. But don't worry—you'll still get the benefits of personalized content, but in a way that's safe and anonymous. It's like enjoying all the perks of being recognized at your favorite coffee shop without ever having to tell them your name. With Web3, you own your keys, you own your identity, and big tech has to adjust to a world where we get to be anonymous again. In this new era of the internet, your data is finally yours—and that's going to make your digital life a whole lot safer and more enjoyable. The future is decentralized, and trust me, it's going to be awesome.

GLOSSARY OF TERMS

AE: Account Executive – A member of the sales team who traditionally hunts for new business or grows existing business within an account. Titles may vary by industry, such as Senior Account Executive, Strategic Account Manager, Regional Account Executive, or Enterprise Account Executive.

AI: Artificial Intelligence – Technology that requires machine learning to simulate human intelligence.

AM: Account Manager – Similar to an Account Executive, but with a stronger focus on managing the success of existing customers. Typically less of a hunter, more of a farmer.

BDR: Business Development Representative (Biz Dev) – An associate-level sales position, usually internal, supporting AEs or AMs through cold calling, making appointments, screening leads, etc.

Bitcoin: The first Cryptocurrency invented by Satoshi Nakamoto (allegedly).

Bleeding Edge Technology - A term used to describe the latest and most advanced technology that is still experimental and not yet widely adopted.

Blockchain: A growing list of records, called blocks, linked using cryptography. Each block contains a cryptographic hash of the previous block, a timestamp, and transaction data, often represented as a Merkle tree.

CRM: Customer Relationship Management – A software platform used to manage contacts, tasks, and sales cycles.

Cryptocurrency: A digital asset designed to function as a medium of exchange. It uses cryptography to secure financial transactions, control the creation of additional units, and verify asset transfers. Cryptocurrencies are decentralized, unlike centralized digital currencies or banking systems.

DevOps: Development Operations – A team of software engineers skilled at managing development operations and quickly gathering coding resources.

Disruptive: An innovation that significantly changes the way people, businesses, and industries operate. It can create new markets, business models, and sectors, and can replace established systems and habits.

Edge: Edge Computing – A distributed computing model that moves computing and data storage closer to the source of the data allowing for devices to process data locally, rather than sending it to a central data center for processing.

ERP: Enterprise Resource Planning – A suite of software that manages various business processes such as internal communications, human resources, accounting, collaborations, data stacks, business systems, supply chain, and relationship management.

Evangelist: A public advocate and expert in a specific type of technology.

FIAT: Inconvertible paper money made legal tender by government decree.

ISV: Independent Software Vendor – A company that develops standalone software for use on an operating system.

Middleware: A software layer that connects applications, databases, and operating systems, allowing them to communicate and share data.

Operating System: Software used to control the operation of computer hardware, such as Linux, Windows, Mac OS, and Android.

Platform: A software system that supports multiple modules.

SaaS: Software as a Service – Software designed to solve business problems through cloud-based solutions.

Scrum: A set of practices used in agile project management that emphasizes daily communication and the flexible reassessment of plans, carried out in short, iterative phases of work.

SE: Sales Engineer – A technically focused member of the sales team who supports AEs at all stages of the sales process.

SLED: State, Local, and Education – Refers to accounts or customers in the state, local government, and education sectors. Often organized as its own sales division.

SOP: Standard Operating Procedure – A prescribed method of performing tasks or processes.

SoW: Scope of Work – A document outlining a project's details and benchmarks.

OKR: Objectives and Key Results – A management methodology that tracks activities related to project objectives across an enterprise and visualizes relationships between objectives.

Unicorn: A tech company valued at over $1 billion USD.

Vaporware: Software that is theoretical and has not been proven to work.

White Paper: A technical explanation of a product, often using examples of past projects to support findings.

ABOUT THE AUTHOR

James L. Toland has served as a Chief Growth Officer of a Silicon Valley stealth startup, SVP of Sales for numerous tech startups, and is currently General Manager of a Canadian enterprise supply chain software company as well as q member of a number of advisory boards as a Go-To-Market Specialist. Father of 2, James enjoys cycling, cooking and rock shows and is currently finishing up his first fiction novel. Toland has been involved in tech since 1996 taking occasional breaks to write and produce independent films in Los Angeles and NYC. Toland, not averse to risk, invests in restaurants, tech startups, independent artist, and real estate. Mr. Toland is a Limited Partner for the Carnegie Hill Group, a technology incubator and venture capital firm in New York City.

Note: Toland is not related to the San Francisco author with a similar name.

Notes:

www.ingramcontent.com/pod-product-compliance
Lightning Source LLC
Chambersburg PA
CBHW040222220526
45473CB00001B/80